Nearly Periodic Matrix Operators For Physics

by

Clifford E. Morgan

authorHOUSE®

AuthorHouse™
1663 Liberty Drive, Suite 200
Bloomington, IN 47403
www.authorhouse.com
Phone: 1-800-839-8640

First published by AuthorHouse 08/09/07

ISBN: 978-1-4343-1445-1 (sc)

Library of Congress Control Number: 2007903989

Printed in the United States of America
Bloomington, Indiana

This book is printed on acid-free paper..

PREFACE

This book grew out of the author's research on mathematical operations in multi-dimensional space. That previous work was more or less completed and copyrighted from August 1997 to April 1999. Nevertheless, the line of research continued to grow since then and some important discoveries have been made.

A class of matrices, which have a particular structure and are periodic and "nearly" periodic, is first briefly described. In particular, nearly periodic matrix operators constructed from suitable 4-vectors turn out to be useful in theoretical physics. The properties of operators of this type are described in Chapters II and III. Rotation and inversion operators based on this type matrix are discussed in Chapter IV. In Chapters V, VI, and VII useful application of these operators in Relativistic Mechanics, Electromagnetic Theory, Quantum Mechanics, and very briefly Classical Mechanics is demonstrated, whereby many of the differential equations of theoretical physics are derived merely by repeated application of the operators to a suitable 4-vector. For example, several important differential equations of E-M Theory (from charge-current continuity to Maxwell's equations) are generated by repeated application of the nearly periodic differential operator on the 4-vector potential of Electromagnetism. Two important results are the construction of a correct generalized Lorentz transformation matrix and derivation of a relativistic general velocity addition rule. Further development of ideas is carried out in Chapters VIII through XI. Important results are summarized and discussed in Chapter XII

Near the end of writing this book, it was found that the relativistic general velocity addition rule yields, because of its cross product nature, the same predictions as General Relativity for perihelion drift, bending of light rays, and the gravitational redshift. This result is astonishing and could well be an error or an odd coincidence. It needs, of course, to be checked by independent workers. Chapter XIII was written to explore the possibility of a flat space gravitational field, or, perhaps gravitation resulting from velocity curvature instead

This was not intended to be a textbook, but a report of research results instead, and a multitude of important topics in physics are not included. In spite of these omissions, however, this book could still be used as a supplement, at least, to a modern theoretical physics text. The main purpose in writing this book was to make available to other physicists and students of physics the unique results and new ideas without compromising the author's claim to prior discovery.

I plan to continue this work so long as it is possible for me to do so. This book will be, I hope, not the only one on this aspect of physics and I expect that it will be revised many times. We physicists should always explore other viewpoints that have merit and endeavor to find and test as many alternative hypotheses as feasible. This must, of course, be limited and tempered by experiment, the ultimate proof of validity.

Finally, it is hoped that the new ideas and discoveries will lead to new and better insights to physical processes and physics itself and inspire creativity in others.

Jerusalem on the Brazos, Texas   Clifford E. Morgan
October 2006

CONTENTS

Chapter I. Introduction

Equation Section (Next)

Operator equations are mathematical representations of physical processes. Almost needless to say, operators are an important part of modern mathematical physics. An operator is a mathematical procedure which is applied to a mathematical function to obtain a new mathematical function. Examples are differentiation of a function with respect to a variable of the function, or something so simple as multiplication of a set of numbers by another set of numbers according to some rule specified by the operator. Operators are usually applied to mathematical functions corresponding to the states of a physical system. And, if skillfully constructed, operator equations parallel the action of a physical process on the state of the system. Operators may be applied repeatedly to a particular function, although they are nearly always applied only once or, at most, twice. While the mathematical manipulations may not have much resemblance to the physical processes, it is sometimes possible to get valid insights to some of the physics involved from the nature of the operators and the outcome of the operation. In this work, the results of a first attempt to use a new approach to the development of operators and operator equations is presented which has the main purpose of gaining new insight into the corresponding physics.

There are a multitude of different operators presently in use, chiefly of course, in quantum mechanics. In this work, attention will be confined to matrix operators, which, in general, are linear and have considerable elegance and which, in particular, belong to a class of matrices obeying a certain type of matrix equation and having similar determinants and characteristic functions. It will be shown that these operators can be applied not only to quantum theory, but also to relativistic mechanics and electromagnetic theory yielding new insight. The latter applications are not traditional, and are apparently not known outside this work.

The author believes that a wide range of viewpoints is a good way to increased understanding in complicated subjects like physics and also to higher levels of creativity, which, unfortunately are greatly lacking, most especially in science and culture these days on this side of the ocean. This research is a continuing work, with much hope that many new applications and discoveries will be made.

Chapter II. Periodic and Nearly Periodic Matrices

Equation Section (Next)

The nearly periodic type matrix was discovered by the author while researching periodic matrices. Some of the properties of periodic matrices will be briefly described in the following, in order to give a better understanding of what is meant by a nearly periodic matrix.

A periodic matrix M of period $n-1$ and any number of dimensions is defined by the equation

$$M^n \equiv pM, \quad (2.1)$$

where p is a power function of the real or complex elements of M and n is the smallest positive integer for which equation (2.1) is valid. Examples of symmetric periodic matrices, A , with period = 1 and different dimensions are

$$A = \begin{pmatrix} a^2 & ab \\ ba & b^2 \end{pmatrix}, \quad \begin{pmatrix} a^2 & ab & ac \\ ba & b^2 & bc \\ ca & cb & c^2 \end{pmatrix}, \quad \begin{pmatrix} a^2 & ab & ac & ad \\ ba & b^2 & bc & bd \\ ca & cb & c^2 & cd \\ da & db & dc & d^2 \end{pmatrix}, \text{ etc, } (2.2)$$

$$A^2 = \left(a^2 + b^2\right)A, \quad \left(a^2 + b^2 + c^2\right)A, \quad \left(a^2 + b^2 + c^2 + d^2\right)A, \text{ etc.} (2.3)$$

Reversed periodic matrices, A_r , of period = 1 and different dimensions are given by

$$A_r = \begin{pmatrix} ab & a^2 \\ b^2 & ba \end{pmatrix}, \quad \begin{pmatrix} ac & ab & a^2 \\ bc & b^2 & ba \\ c^2 & cb & ca \end{pmatrix}, \quad \begin{pmatrix} ad & ac & ab & a^2 \\ bd & bc & b^2 & ba \\ cd & c^2 & cb & ca \\ d^2 & dc & db & da \end{pmatrix}, \text{ etc.} (2.4)$$

$$A_r^2 = (2ab)A_r, \quad \left(2ac + b^2\right)A_r, \quad 2(ad + bc)A_r, \text{ etc.} (2.5)$$

Periodic matrices of period =1 which are exactly equal to their squares are also called idempotent matrices. Examples of antisymmetric matrices, X, are

3

$$X = \begin{pmatrix} 0 & a \\ -a & 0 \end{pmatrix}, \quad \begin{pmatrix} 0 & -c & b \\ c & 0 & -a \\ -b & a & 0 \end{pmatrix}, \quad \begin{pmatrix} 0 & a & b & c \\ -a & 0 & -c & b \\ -b & c & 0 & -a \\ -c & -b & a & 0 \end{pmatrix}, \text{ etc.(2.6)}$$

$$X^2 = -a^2 I, \quad A - \left(a^2 + b^2 + c^2\right)I, \quad -\left(a^2 + b^2 + c^2\right)I, \text{ etc., (2.7)}$$

where I is the identity matrix. Note that the even dimensional X matrices have period = 2, while the odd dimensional X matrices are not periodic. The elements a, b, c, etc. for the moment are real or complex numbers; however, simple differential operators corresponding to Cartesian coordinates may be substituted. There is one very important caveat for substitution of operators in any of these matrices and that is the operators must commute or anticommute for some of the equations, such as (2.3), (2.5), and(2.7) to be valid. Note that determinants of all A matrices equal zero, determinants of 2^n even dimensional X matrices equal a^2, $a^2 + b^2 + c^2$, etc., and determinants of all odd dimensional X matrices equal zero. X and A matrices, both of an odd number of dimensions, commute, thus

$$AX = XA = 0. \qquad (2.8)$$

Incidentally, the odd dimensional X matrices, when multiplied from the left on a vector of an appropriate odd number of dimensions, yield the positive cross product of that vector with the vector from which X was constructed. These cross products are complete (form a group) only for X matrices of dimension $n = 2^m - 1$, where m is any positive integer up to at least five. These particular X matrices will, naturally, be called cross matrices. Finally, one should note that the famous Pauli spin matrices are periodic matrices with period = 2.

A nearly periodic matrix, N, of any number of dimensions, and of quasi period = 1 is defined by the equation

$$N^2 = 2p_1 N + p_2^2 I, \qquad (2.9)$$

where p_1 and p_2 are power functions of the elements of N. (This kind of matrix will sometimes be referred to as an "NP" matrix in the following.) Nearly periodic matrices are restricted to an even number of dimensions given by $n = 2^i$, where i is any positive integer. The NP matrix N can be converted to a periodic matrix $N - p_1 I$ of period = 2 by simple algebraic manipulation. Thus, rearranging equation (2.5), one gets

$$(N - p_1 I)^2 = (p_1^2 + p_2^2)I \ (2.10) \ and$$

$$(N - p_1 I)^3 = (p_1^2 + p_2^2)(N - p_1 I) \ (2.11)$$

It should be carefully noted that the square root of equation (2.10) cannot be taken, i.e.

$$N - p_1 I \neq \left(\sqrt{p_1^2 + p_2^2} \right) I .$$

The author has made a fairly extensive study of the geometric and other mathematical properties of both periodic and NP matrices (see Reference 1 - Clifford E. Morgan, *"A Method for Constructing the Vector or Cross Product for Multi-Dimensional Vectors"*, U. S. Army Science and Technology Center, Mainz-Kastel, Germany, copyright August 1997, *"Some Geometrical and Possible Physical Properties of Multidimensional Spaces"*, Mainz-Kastel, Germany, copyright January 1998, and *"Cross Product Operators and Interesting 4-D Geometries"*, Frankfurt am Main, Germany, copyright April 1999). The properties of these matrices indicate that they may be made the basis for important processes in physics. The theme of the next five sections of this work will be a demonstration of this possibility.

Dirac matrices are important examples of matrices with NP character. The representation of the Dirac equation given in Bjorken and Drell (Reference 2 - J. D. Bjorken and S.D. Drell, *Relativistic Quantum Mechanics*, McGraw-Hill (1964), p. 17) is shown here in matrix operator form as

$$D_i \overline{\psi} = \begin{pmatrix} d-p & 0 & c & a-ib \\ 0 & d-p & a+ib & -c \\ c & a-ib & -d-p & 0 \\ a+ib & -c & 0 & -d-p \end{pmatrix} \begin{pmatrix} \psi_0 \\ \psi_1 \\ \psi_2 \\ \psi_3 \end{pmatrix} = 0 \qquad (2.12)$$

where we have taken the liberty of replacing the differential operators in the original Dirac equation with the symbols d, a, b, and c in order to show more clearly the nearly periodic character of the Dirac type matrix. The matrix D_i obeys the equations

$$D_i^2 = -2pD_i + (a^2 + b^2 + c^2 + d^2 - p^2)I$$
$$(D_i + pI)^2 = (a^2 + b^2 + c^2 + d^2)I$$
,(2.13)

and its determinant is equal to $(a^2 + b^2 + c^2 + d^2 - p^2)^2$. To show that a different representation of the matrix can change the matrix equations in interesting ways, another Dirac type matrix taken from the representation in Schiff (Reference 3 - L. I. Schiff, *Quantum Mechanics,*.McGraw-Hill, New York (1955), p. 327) is exhibited below with attendant matrix equations

$$D_i = \begin{pmatrix} d-p & -a-ib & c & 0 \\ -a+ib & d+p & 0 & -c \\ c & 0 & d+p & -a-ib \\ 0 & -c & -a+ib & d-p \end{pmatrix} \quad (2.14)$$

And

$$D_i^2 = -2dD_i + (a^2 + b^2 + c^2 - d^2 + p^2)I$$
$$(D_i - dI)^2 = (a^2 + b^2 + c^2 + p^2)I$$
. (2.15)

Its determinant is equal to $\left(a^2 + b^2 + c^2 - d^2 + p^2\right)^2$ Finaly, an example of yet another representation of the Dirac type matrix, again from Bjorken and Drell (Reference 2 - p. 6), is shown with attendant matrix equations,

$$D_i = \begin{pmatrix} d+p & 0 & c & a-ib \\ 0 & d+p & a+ib & -c \\ c & a-ib & d-p & 0 \\ a+ib & -c & 0 & d-p \end{pmatrix} \quad (2.16) \; and$$

$$D_i^2 = 2dD_i + (a^2 + b^2 + c^2 - d^2 + p^2)I$$
$$(D_i - dI)^2 = (a^2 + b^2 + c^2 + p^2)I$$

Its determinant is equal to $\left(a^2 + b^2 + c^2 - d^2 + p^2\right)^2$. Although the attendant equations are somewhat altered, the various representations of the Dirac type matrix are all matrices with NP character and all have similar determinants, of course. In the original Dirac equation, the matrix elements are, of course, simple differential operators or numbers and are given by

$$a = i\hbar \frac{\partial}{\partial x}, \quad b = i\hbar \frac{\partial}{\partial y}, \quad c = i\hbar \frac{\partial}{\partial z}, \quad d = \frac{i\hbar}{c} \frac{\partial}{\partial t}, \quad p = mc. \quad (2.17)$$

Dirac matrices exhibit an almost incredible amount of ensnarlment of the elements. Dirac matrices, with various degrees of ensnarlment, can be easily constructed by introducing positive and negative elements, p, in addition to the d's on the principal diagonal and then requiring the matrix to be nearly periodic, thus

$$D_i = \begin{pmatrix} d+p & a-ic & -b & 0 \\ a+ic & d-p & 0 & b \\ -b & 0 & d-p & a+ic \\ 0 & b & a-ic & d+p \end{pmatrix} \quad and$$

$$D_i = \begin{pmatrix} d+p & c-ib & 0 & a \\ c+ib & d-p & -a & 0 \\ 0 & a & d+p & c+ib \\ -a & 0 & c-ib & d-p \end{pmatrix},$$

where judicious use of the imaginary number i is made to avoid cross terms when squaring the matrix. Both of these matrices represent nearly maximal ensnarlment of components for four dimensional matrices.

With the periodic and nearly periodic NP matrices more or less defined, the properties of periodic matrices and the less important properties of NP matrices will not be further discussed in the following text. For more information on periodic and NP matrices, see Reference 1, above.

Chapter III. The □ Operator - An Especially Elegant Type of Nearly Periodic Matrix

Equation Section (Next)

The especially elegant NP matrix, which will be designated by the symbol N, is formed by adding another row and column and nonzero diagonal elements to a cross product matrix. The new elements are ordered in a simple way which is made clear by a couple of examples. N Matrices are, therefore, even dimensional matrices of dimension $= 2^n$ where n is a positive integer. The effect of the added row and column is that, when a 4-vector is multiplied on the left by N, it yields a 4-vector, the fourth component of which is a dot product of the vector, $^4\overline{d} = d\hat{i} + a\hat{i} + b\hat{j} + c\hat{k}$ with the vector operated on, and the space components are a type of cross product of these two vectors. There are four basic types of the four dimensional form of this matrix, not counting negatives and which will be designated by $\square_1$, $\square_1^*$, $\square_2$, and $\square_2^*$, thus

$$\square_1 = \begin{pmatrix} d & a & b & c \\ -a & d & -c & b \\ -b & c & d & -a \\ -c & -b & a & d \end{pmatrix}, \quad \square_1^* = \begin{pmatrix} -d & a & b & c \\ -a & -d & -c & b \\ -b & c & -d & -a \\ -c & -b & a & -d \end{pmatrix},$$

$$\square_2 = \begin{pmatrix} d & a & b & c \\ -a & d & c & -b \\ -b & -c & d & a \\ -c & b & -a & d \end{pmatrix}, \quad \square_2^* = \begin{pmatrix} -d & a & b & c \\ -a & -d & c & -b \\ -b & -c & -d & a \\ -c & b & -a & -d \end{pmatrix} .(3.1)$$

When multiplied from the left, the two $\square_1$ matrices yield positive cross products, whilst the two $\square_2$ matrices yield negative cross products with vectors. The starred matrices are conjugates of their partners. The standard four dimensional matrix is chosen to be

8

$\square_1$ shown above and the standard eight dimensional matrix is chosen to be

$$\square_1 = \begin{pmatrix} d & a & b & c & h & e & f & g \\ -a & d & -c & b & -e & h & g & -f \\ -b & c & d & -a & -f & -g & h & e \\ -c & -b & a & d & -g & f & -e & h \\ -h & e & f & g & d & -a & -b & -c \\ -e & -h & g & -f & a & d & c & -b \\ -f & -g & -h & e & b & -c & d & a \\ -g & f & -e & -h & c & b & -a & d \end{pmatrix} \quad (3.2)$$

The determinant of the 4-D forms of the N matrix is

$$det\,\square = \left(d^2 + a^2 + b^2 + c^2\right)^2. \square$$

and the determinant of the 8-D forms of the N matrix is

$$det\,\square = \left(d^2 + a^2 + b^2 + c^2 + h^2 + e^2 + f^2 + g^2\right)^4$$

The latter was determined by hand calculation of all 40,320 terms of the determinant (most of which cancelled each other, Gott sei dank) and, therefore, should be checked by independent workers.

Higher than eight dimensional matrices of the NP type cannot be constructed using only simple elements such as real or complex numbers or simple differential operators. In these higher dimensional matrices the elements must themselves be matrices or objects equally complex as matrices (i.e. with internal structure). Some very important properties of the four dimensional matrices are listed here below.

The N type matrices have some remarkable properties apart from being nearly periodic, three of which will now be briefly discussed. The first is that they are formed from a Latin square arranged with the same element for each of the main diagonal elements. As pointed out later, this arrangement permits one to get the proper Maxwell equations when operating on a 4-vector potential and does not mix up the field components as a more arbitrary Latin square arrangement would. The second property is that two $\square$ matrices, having no common elements, when multiplied together yield another $\square$ matrix. For example,

9

$$\begin{pmatrix} d & a & b & c \\ -a & d & -c & b \\ -b & c & d & -a \\ -c & -b & a & d \end{pmatrix} \begin{pmatrix} h & e & f & g \\ -e & h & -g & f \\ -f & g & h & -e \\ -g & -f & e & h \end{pmatrix} = \begin{pmatrix} s & u & v & w \\ -u & s & -w & v \\ -v & w & s & -u \\ -w & -v & u & s \end{pmatrix}$$

(3.3) where

$$s = dh - ae - bf - cg$$
$$u = de + ah - bg + cf$$
$$v = df + ag + bh - ce$$
$$w = dg - af + be + ch$$

The third property is that the vector formed by multiplying any arbitrary vector with N from the left or right yields a new vector, each component of which consists of a sum of terms, that when squared, yields a product for which all the cross terms add to zero. To clarify this statement, the following example is presented:

$$\begin{pmatrix} d & a & b & c \\ -a & d & -c & b \\ -b & c & d & -a \\ -c & -b & a & d \end{pmatrix} \begin{pmatrix} s \\ x \\ y \\ z \end{pmatrix} = (ds + ax + by + cz)\hat{l} + (-as + dx + bz - cy)\hat{i}$$

$$+(-bs + dy + cx - az)\hat{j} + (-cs + dz + ay - bx)\hat{k}$$

From which it can be seen that

$$\left[\begin{array}{l} (ds + ax + by + cz)\hat{l} + (-as + dx + bz - cy)\hat{i} \\ +(-bs + dy + cx - az)\hat{j} + (-cs + dz + ay - bx)\hat{k} \end{array} \right]^2 =$$

$$(ds + ax + by + cz)^2 + (-as + dx + bz - cy)^2 + (-bs + dy + cx - az)^2 + (-cs + dz + ay - bx)^2 =$$
$$(d^2 + a^2 + b^2 + c^2)(s^2 + x^2 + y^2 + z^2)$$

.

This feature permits us to do what I will call unsnarling and unsquaring operations on the vectors that are obtained with our N matrix operators. Even more remarkable is that the vector obtained by operating with an N matrix on an arbitrary vector can

10

also be operated on with another otherwise arbitrary N matrix to give yet another vector with similar properties, thus

$$
\begin{pmatrix} h & e & f & g \\ -e & h & -g & f \\ -f & g & h & -e \\ -g & -f & e & h \end{pmatrix} \begin{pmatrix} (ds + ax + by + cz) \\ (-as + dx + bz - cy) \\ (-bs + dy + cx - az) \\ (-cs + dz + ay - bx) \end{pmatrix} =
$$

$$
+ \begin{bmatrix} h(ds + ax + by + cz) + e(-as + dx + bz - cy) + \\ f(-bs + dy + cx - az) + g(-cs + dz + ay - bx) \end{bmatrix} \hat{i}
$$

$$
+ \begin{bmatrix} -e(ds + ax + by + cz) + h(-as + dx + bz - cy) + \\ -g(-bs + dy + cx - az) + f(-cs + dz + ay - bx) \end{bmatrix} \hat{i} \ .
$$

$$
+ \begin{bmatrix} -f(ds + ax + by + cz) + g(-as + dx + bz - cy) + \\ +h(-bs + dy + cx - az) - e(-cs + dz + ay - bx) \end{bmatrix} \hat{j}
$$

$$
+ \begin{bmatrix} -g(ds + ax + by + cz) - f(-as + dx + bz - cy) + \\ e(-bs + dy + cx - az) + h(-cs + dz + ay - bx) \end{bmatrix} \hat{k}
$$

This new vector, $\overline{w}$, when squared yields

$$
\overline{w}^2 = \left(e^2 + f^2 + g^2 + h^2\right)\left(a^2 + b^2 + c^2 + d^2\right)\left(s^2 + x^2 + y^2 + z^2\right)
$$

Repeated operation with □ yields similar results and so on. This allows one to unsnarl vectors formed by operation with □ matrices in the following manner. The nature of $\overline{w}^{-2}$, for example, is such that it is the same as if $\overline{w}$ were (the magnitude of a vector) times (the magnitude of a vector) times (a vector), i.e., if we did not know better, $\overline{w}$ is one of three possible vectors, namely

$$
1.) \quad \overline{w} = |\overline{e}||\overline{a}|\left(s\hat{i} + x\hat{i} + y\hat{j} + z\hat{k}\right)
$$

$$
2.) \quad \overline{w} = |\overline{e}||\overline{s}|\left(d\hat{i} + a\hat{i} + b\hat{j} + c\hat{k}\right) \quad \text{or}
$$

11

$$3.)\quad \overline{w} = \left|\overline{a}\right|\left|\overline{s}\right|\left(h\hat{l} + e\hat{i} + f\hat{j} + g\hat{k}\right) \text{ where}$$

$$\left|\overline{a}\right| = \sqrt{d^2 + a^2 + b^2 + c^2}\ ,\ \left|\overline{e}\right| = \sqrt{h^2 + e^2 + f^2 + g^2}\ ,$$

$$\text{and } \left|\overline{s}\right| = \sqrt{s^2 + x^2 + y^2 + z^2}\ .$$

There are two ways, at least, to unsnarl such a vector. One unsnarling operation consists in squaring the vector, deciding which is the correct vector (from a choice like 1.), 2.), or 3.) above), and then unsquaring to get one of the vectors. The second method is used when this type of vector has been used to make a scalar product with another vector. It is more applicable to arguments of wavefunctions and will be used more fully in Chapter VII, *"Application of $\square$ Matrices in Quantum Mechanics"*. It is usually fairly obvious which the correct choice is (especially if we know it beforehand). Actually this choice can be made on the basis of what would be expected using classical physics reasoning or the classical expectation value. In this respect, this operation fits in with the modern principle of Quantum Mechanics that basically says that we can use any kind of mathematics, reasoning, or model (no matter how insane) as long as the calculations give the correct answer in agreement with experiment to high precision. This apparently holds true even though we cannot really say if the model actually has any further connection with reality. Therefore, the author justifies the unsnarling procedure on the basis that it gives the correct result to high precision.

These properties of the $\square$ matrices, which apply to $\square$ matrices of dimension $n = 2$ up to at least $n = 8$ (and maybe $n = 16$ and $n = 32$), allow us to encode (in a sense) vectors of even dimensions (in this range of dimensions at least), such as the vector $\overline{w}$ given as an example above, with information from a huge (perhaps infinite) number of vectors. And not only that, but it is also possible to recover any one of the individual vectors by the unsnarling operation, described above. If the vectors are now taken to be Quantum Mechanical state vectors, we have then a mechanism for "encoding" wavefunctions with information about a great many possible states of a system by means of operators (the appropriate $\square$ matrices) and recovering information about particular states by an operation that involves squaring, unsnarling and, perhaps, unsquaring. Such properties should, therefore be very important in Quantum Mechanics. If we incorporate a 4-vector, such as $\bar{w}$ above, in a suitable form as the argument of a wavefunction, it would be possible, then to write a wave function which has all of the possible states existing in a phase space of very large dimension. In Chapter VII, the argument for the wave function of a single free particle is derived from a vector which is such a combination of the velocity and position vector of the particle, namely the Lorentz-transformed position vector dotted into a propagation vector, $\bar{k}$.

In all fairness, there is (not counting negatives of these matrices) another 4-D matrix, besides the ☐ matrix, which has the property that when multiplied into an arbitrary 4-vector it yields a vector, which when squared, has all of the cross terms also summing to zero. This matrix is

$$H = \begin{pmatrix} -d & a & b & c \\ a & d & -c & b \\ b & c & d & -a \\ c & -b & a & d \end{pmatrix} \qquad (3.4)$$

Whereby,

$$H \begin{pmatrix} s \\ x \\ y \\ z \end{pmatrix} = \begin{pmatrix} -ds + ax + by + cz \\ as + dx - cy + bz \\ bs + cx + dy - az \\ cs - bx + ay + dz \end{pmatrix} \qquad (3.5)$$

The square of this new vector is

$$\left(H \begin{pmatrix} s \\ x \\ y \\ z \end{pmatrix} \right)^2 = \left(d^2 + a^2 + b^2 + c^2 \right)\left(s^2 + x^2 + y^2 + z^2 \right) \quad (3.6)$$

Therefore the H operator can also be used similarly to encode vectors with information.

In normal physics, the most useful of these NP matrices are the four dimensional ones with general real or complex numbers or simple Cartesian differential operators, e.g. $\dfrac{i}{c}\dfrac{\partial}{\partial t}$, $\dfrac{\partial}{\partial x}$, $\dfrac{\partial}{\partial y}$, or $\dfrac{\partial}{\partial z}$, as elements. These matrices obey the equations

$$\square_1^2 = 2d\square_1 - (a^2 + b^2 + c^2 + d^2)I$$

$$(\square_1 - dI)^2 = -(a^2 + b^2 + c^2)I$$

$$\square_2^2 = 2d\square_2 - (a^2 + b^2 + c^2 + d^2)I$$

$$(\square_2 - dI)^2 = -(a^2 + b^2 + c^2)I$$

$$\square_1^{*2} = -2d\square_1^* - (a^2 + b^2 + c^2 + d^2)I \qquad \text{(3.7) and}$$

$$(\square_1^* + dI)^2 = -(a^2 + b^2 + c^2)I$$

$$\square_2^{*2} = -2d\square_2^* - (a^2 + b^2 + c^2 + d^2)I$$

$$(\square_2^* + dI)^2 = -(a^2 + b^2 + c^2)I$$

$$\begin{aligned} \square_1\square_1^* &= \square_1^*\square_1 = -(a^2 + b^2 + c^2 + d^2)I \\ \square_2\square_2^* &= \square_2^*\square_2 = -(a^2 + b^2 + c^2 + d^2)I \end{aligned} \qquad \text{(3.8)}$$

Multiplication across types of matrix yields rather odd results, for example

$$\square_1\square_2 = 2d\begin{pmatrix} d & a & b & c \\ -a & 0 & 0 & 0 \\ -b & 0 & 0 & 0 \\ -c & 0 & 0 & 0 \end{pmatrix} - 2\begin{pmatrix} 0 & 0 & 0 & 0 \\ 0 & a^2 & ab & ac \\ 0 & ba & b^2 & bc \\ 0 & ca & cb & c^2 \end{pmatrix} + (a^2+b^2+c^2+d^2)\begin{pmatrix} -1 & 0 & 0 & 0 \\ 0 & 1 & 0 & 0 \\ 0 & 0 & 1 & 0 \\ 0 & 0 & 0 & 1 \end{pmatrix} \text{ (3.9)}$$

The $\square$ matrices are tensors, but, as was seen, they can also be associated with a four dimensional vector, such as

$$d\hat{l} + a\hat{i} + b\hat{j} + c\hat{k}, \qquad (3.10)$$

where $\hat{l}$ is the unit vector in the fourth or time dimension and $\hat{i}$, $\hat{j}$, and $\hat{k}$ are the usual three dimensional unit vectors for the x, y, and z directions respectively. The reader will recall that the determinant of any 4-dimensional $\square$ matrix is equal to $(d^2 + a^2 + b^2 + c^2)^2$; consequently, the eigenvalues are easily found. The eigenvalues, λ_i and eigenvectors, Λ_i, of the $\square_1$ matrix, for example, are listed here:

$$\lambda_0 = d + i\sqrt{a^2 + b^2 + c^2} \qquad \Lambda_0 = \left(i\sqrt{a^2 + b^2 + c^2}, \quad a, \quad b, \quad c\right)$$

$$\lambda_1 = d - i\sqrt{a^2 + b^2 + c^2} \qquad \Lambda_1 = \left(-i\sqrt{a^2 + b^2 + c^2}, \quad a, \quad b, \quad c\right)$$

$$\lambda_2 = -d + i\sqrt{a^2 + b^2 + c^2} \qquad \Lambda_2 = \left(i\sqrt{a^2 + b^2 + c^2}, \quad -a, \quad -b, \quad -c\right)$$

$$\lambda_3 = -d - i\sqrt{a^2 + b^2 + c^2} \qquad \Lambda_3 = \left(-i\sqrt{a^2 + b^2 + c^2}, \quad -a, \quad -b, \quad -c\right)$$

The eigenvalues are all real if a, b, and c are pure imaginary and d is pure real, but not if a, b, and c are pure real and d is pure real! Also, if a, b, and c are pure real and d pure imaginary, then the eigenvalues are all pure imaginary. The N matrices are therefore not restricted to being Hermitian matrices and the eigenvalues can be not only pure real, but also pure imaginary. This fits in with, and is apparently required by our odd Minkowski space. The N matrix approach is therefore less restrictive than the usual approach.

A digression is made here to point out ahead of time some peculiarities of using ☐ matrices as operators that we will encounter later. In Minkowski space d is taken as pure imaginary and a, b, and c as pure real *or a, b, and c, are taken as pure imaginary and d as pure real*, in order for the space to be Euclidian. The convention in this work will be to use an odd kind of Minkowski space in which vectors having pure imaginary time components and pure real space components will be regarded as distinct and separate vectors from those having pure real time components and pure imaginary space components. Both types of vectors will occur in the same equation, but, as will be seen are fairly easily distinguished. In spite of these difficulties, there are advantages to this approach, as, for example, the ease with which Lorentz transforms can be made, one operation on the E-M fields (as it will be seen) yields the entire set of Maxwell's equations, and the useful application of one type of operator to a number of fields of theoretical physics. This is not to mention the possibility of new insights and deeper understanding that may be garnered. It begins to appear that special relativity should, perhaps, have built into it such a space as this Minkowski space.

Returning to the discussion of especially elegant ($\square$) matrices, we note that symbols other than $\square$ will be used in the following for special matrices such as that for the Lorentz transform, $\mathcal{L}$; the differential operator matrix, D ; or the Dirac matrix, D_i. The Lorentz transformation matrix is given by

$$\mathcal{L}_1 = -i \begin{pmatrix} i\gamma & \gamma\beta_x & \gamma\beta_y & \gamma\beta_z \\ -\gamma\beta_x & i\gamma & -\gamma\beta_z & \gamma\beta_y \\ -\gamma\beta_y & \gamma\beta_z & i\gamma & -\gamma\beta_x \\ -\gamma\beta_z & -\gamma\beta_y & \gamma\beta_x & i\gamma \end{pmatrix} \qquad (3.11)$$

and is associated with the 4-vector velocity $i\gamma c\hat{1} + \gamma c\bar{\beta}$. The differential operator matrix is given by

$$D_1^* = \begin{pmatrix} \dfrac{i}{c}\dfrac{\partial}{\partial t} & \dfrac{\partial}{\partial x} & \dfrac{\partial}{\partial y} & \dfrac{\partial}{\partial z} \\[6pt] \dfrac{\partial}{\partial x} & \dfrac{i}{c}\dfrac{\partial}{\partial t} & \dfrac{\partial}{\partial z} & \dfrac{\partial}{\partial y} \\[6pt] \dfrac{\partial}{\partial y} & \dfrac{\partial}{\partial z} & \dfrac{i}{c}\dfrac{\partial}{\partial t} & \dfrac{\partial}{\partial x} \\[6pt] \dfrac{\partial}{\partial z} & \dfrac{\partial}{\partial y} & \dfrac{\partial}{\partial x} & \dfrac{i}{c}\dfrac{\partial}{\partial t} \end{pmatrix} \qquad (3.12)$$

D_1^* is associated with the four dimensional differential operator vector

$$-\frac{i}{c}\frac{\partial}{\partial t}\hat{1} + \nabla \qquad (3.13)$$

A word of caution is necessary when the elements of the N matrices are changed to differentials, namely, a Cartesian coordinate yields the correct results. Thus, in using spherical coordinates, one must employ matrix elements as shown in the matrix,

$$\begin{pmatrix} \dfrac{i}{c}\dfrac{\partial}{\partial t} & \dfrac{\partial}{\partial r}+\dfrac{2}{r} & \dfrac{1}{r}\dfrac{\partial}{\partial\theta}+\dfrac{\cot\theta}{r} & \dfrac{1}{r\sin\theta}\dfrac{\partial}{\partial\phi} \\[2ex] -\dfrac{\partial}{\partial r} & \dfrac{i}{c}\dfrac{\partial}{\partial t} & -\dfrac{1}{r\sin\theta}\dfrac{\partial}{\partial\phi} & \dfrac{1}{r}\dfrac{\partial}{\partial\theta}+\dfrac{\cot\theta}{r} \\[2ex] -\dfrac{1}{r}\dfrac{\partial}{\partial\theta} & \dfrac{1}{r\sin\theta}\dfrac{\partial}{\partial\phi} & \dfrac{i}{c}\dfrac{\partial}{\partial t} & -\dfrac{\partial}{\partial r}-\dfrac{1}{r} \\[2ex] -\dfrac{1}{r\sin\theta}\dfrac{\partial}{\partial\phi} & -\dfrac{1}{r}\dfrac{\partial}{\partial\theta} & \dfrac{\partial}{\partial r}+\dfrac{1}{r} & \dfrac{i}{c}\dfrac{\partial}{\partial t} \end{pmatrix}, \text{ to get}$$

the correct results. It is obviously easier to use Cartesian coordinates, and all the equations in this book will be in Cartesian coordinates, unless otherwise specified.

Chapter IV. Rotations and Inversions in Multi-Dimensional Spaces

Equation Section (Next)

Transformations, such as rotations, in spaces of dimension greater than three, are not very intuitive and have many ambiguities. Because it has been derived from a cross product matrix (and therefore, can be related to an infinitesimal rotation), it is reasonable to use the ☐ matrix as a guide to developing a rotation matrix. Some very basic requirements for a rotation operator are as follows. If the rotation operator for a given rotation angle, ξ, is applied n times in succession about the same axis, it should result in a rotation by an angle of $n\xi$. Further, it is required that a rotation by 180 degrees about any axis should give negative, or opposite directions, of all the remaining coordinate axes; and a rotation of any angle should not result in any change in the component of the vector corresponding to the axis of rotation. This latter requirement seems especially correct intuitively for Euclidian type spaces. At first we will not require that the determinant of the rotation matrix to be equal to +1, as this is too restrictive. For rotation about the x axis then

$$R_x = \begin{pmatrix} e^{i\alpha} & 0 & 0 & 0 \\ 0 & 1 & 0 & 0 \\ 0 & 0 & \cos a & -\sin\alpha \\ 0 & 0 & \sin\alpha & \cos\alpha \end{pmatrix}, \qquad (4.1)$$

which is consistent with our rules. However, the determinant of R_x is $e^{i\alpha}$, that is, it is not necessarily equal to +1, as befitting a normal orthogonal matrix. If R_x is applied to a vector, it is readily seen that this result is apparently a consequence of our odd Minkowski space,

$$R_x\left(ict\hat{l}+\bar{r}\right)=icte^{i\alpha}\hat{l}+x\hat{i}+\left(y\cos\alpha-z\sin\alpha\right)\hat{j}+\left(z\cos\alpha+y\sin\alpha\right)\hat{k}.$$

18

Hence, if we square this equation, we do not recover $-c^2t^2 + r^2$ as we would expect if R_x were orthogonal, instead we get

$$\left(R_x\left(ic\hat{t} + \bar{r}\right)\right)^2 = -c^2t^2 e^{i2\alpha} + r^2.$$

One way of avoiding this problem is to multiply R_x by its complex conjugate to get the correct result. (Another way would be replacement of $e^{i\alpha}$ in the time element with another, but perhaps very unusual, function of the rotation angle.) Our rotation matrices are orthogonal in this sense. Continuing, we get for rotations about y, z, and ct,

$$R_y = \begin{pmatrix} e^{i\beta} & 0 & 0 & 0 \\ 0 & \cos\beta & 0 & \sin\beta \\ 0 & 0 & 1 & 0 \\ 0 & -\sin\beta & & \cos\beta \end{pmatrix} \qquad (4.2)$$

$$R_z = \begin{pmatrix} e^{i\gamma} & 0 & 0 & 0 \\ 0 & \cos\gamma & -\sin\gamma & 0 \\ 0 & \sin\gamma & \cos\gamma & 0 \\ 0 & 0 & 0 & 1 \end{pmatrix} \qquad (4.3)$$

$$R_t = \begin{pmatrix} 1 & 0 & 0 & 0 \\ 0 & e^{i\delta} & 0 & 0 \\ 0 & 0 & e^{i\delta} & 0 \\ 0 & 0 & 0 & e^{i\delta} \end{pmatrix}. \qquad (4.4)$$

It is readily seen that none of these rotations can produce a fourth component of a vector if none exists beforehand. We list now a set of 4 dimensional inversions derived from the N matrix.

$$i\gamma_y = X = \begin{pmatrix} 0 & 1 & 0 & 0 \\ -1 & 0 & 0 & 0 \\ 0 & 0 & 0 & -1 \\ 0 & 0 & 1 & 0 \end{pmatrix}$$

$$\delta_0 = Y = \begin{pmatrix} 0 & 0 & 1 & 0 \\ 0 & 0 & 0 & 1 \\ -1 & 0 & 0 & 0 \\ 0 & -1 & 0 & 0 \end{pmatrix}$$

$$i\beta_y = Z = \begin{pmatrix} 0 & 0 & 0 & 1 \\ 0 & 0 & -1 & 0 \\ 0 & 1 & 0 & 0 \\ -1 & 0 & 0 & 0 \end{pmatrix}$$

$$-i\alpha_0 = T = \begin{pmatrix} -i & 0 & 0 & 0 \\ 0 & -i & 0 & 0 \\ 0 & 0 & -i & 0 \\ 0 & 0 & 0 & -i \end{pmatrix}.$$

(The 4-D matrices, are defined here and on pages 40 - 42 below where the group is further expanded and their multiplication table provided.)

Incidentally, it is also clear that inversions, such as the above, can produce a fourth component of a vector, even if no fourth component exists beforehand in the vector. In the case of the Lorentz transform of a 3-D force as we shall see later, the fourth component of a Lorentz transformed force will be the work done by the force, e.g.

$$\pounds_7 \begin{pmatrix} 0 \\ \overline{F} \end{pmatrix} = \begin{pmatrix} -i\gamma\overline{\beta}\square\overline{F}\hat{i} \\ \gamma\overline{F} - i\gamma\overline{\beta}\times\overline{F} \end{pmatrix} = \begin{pmatrix} -\dfrac{i\gamma}{c}\dfrac{\partial\varepsilon}{\partial t}\hat{i} \\ \gamma\overline{F} - i\gamma\overline{\beta}\times\overline{F} \end{pmatrix}.$$

Where, clearly $c\overline{\beta}\square\overline{F} = \dfrac{\partial\varepsilon}{\partial t}$ is the work done by $\overline{F}$.

Consequently, if the work done by a force is, in general, not a true force, then this result is still consistent with the fact that a fourth dimensional component of a force is never directly observed.

Consider now rotations in higher dimensions. Eight dimensional matrices suffice for all rotations up to eight dimensions and sixteen dimensional matrices suffice likewise up to sixteen dimensions, and so on. All the 8-dimensional rotation matrices are shown here, thus

$$
R_t = \begin{pmatrix}
1 & 0 & 0 & 0 & 0 & 0 & 0 & 0 \\
0 & e^{i\delta} & 0 & 0 & 0 & 0 & 0 & 0 \\
0 & 0 & e^{i\delta} & 0 & 0 & 0 & 0 & 0 \\
0 & 0 & 0 & e^{i\delta} & 0 & 0 & 0 & 0 \\
0 & 0 & 0 & 0 & e^{i\delta} & 0 & 0 & 0 \\
0 & 0 & 0 & 0 & 0 & e^{i\delta} & 0 & 0 \\
0 & 0 & 0 & 0 & 0 & 0 & e^{i\delta} & 0 \\
0 & 0 & 0 & 0 & 0 & 0 & 0 & e^{i\delta}
\end{pmatrix}
\quad (4.5)
$$

$$
R_x = \begin{pmatrix}
e^{i\alpha} & 0 & 0 & 0 & 0 & 0 & 0 & 0 \\
0 & 1 & 0 & 0 & 0 & 0 & 0 & 0 \\
0 & 0 & \cos\alpha & -\sin\alpha & 0 & 0 & 0 & 0 \\
0 & 0 & \sin\alpha & \cos\alpha & 0 & 0 & 0 & 0 \\
0 & 0 & 0 & 0 & \cos\alpha & -\sin\alpha & 0 & 0 \\
0 & 0 & 0 & 0 & \sin\alpha & \cos\alpha & 0 & 0 \\
0 & 0 & 0 & 0 & 0 & 0 & \cos\alpha & \sin\alpha \\
0 & 0 & 0 & 0 & 0 & 0 & -\sin\alpha & \cos\alpha
\end{pmatrix}
\quad (4.6)
$$

$$
R_y = \begin{pmatrix}
e^{i\beta} & 0 & 0 & 0 & 0 & 0 & 0 & 0 \\
0 & \cos\beta & 0 & \sin\beta & 0 & 0 & 0 & 0 \\
0 & 0 & 1 & 0 & 0 & 0 & 0 & 0 \\
0 & -\sin\beta & 0 & \cos\beta & 0 & 0 & 0 & 0 \\
0 & 0 & 0 & 0 & \cos\beta & 0 & -\sin\beta & 0 \\
0 & 0 & 0 & 0 & 0 & \cos\beta & 0 & -\sin\beta \\
0 & 0 & 0 & 0 & \sin\beta & 0 & \cos\beta & 0 \\
0 & 0 & 0 & 0 & 0 & \sin\beta & 0 & \cos\beta
\end{pmatrix}
$$
$$(4.7)$$

$$R_z = \begin{pmatrix} e^{i\gamma} & 0 & 0 & 0 & 0 & 0 & 0 & 0 \\ 0 & \cos\gamma & -\sin\gamma & 0 & 0 & 0 & 0 & 0 \\ 0 & \sin\gamma & \cos\gamma & 0 & 0 & 0 & 0 & 0 \\ 0 & 0 & 0 & 1 & 0 & 0 & 0 & 0 \\ 0 & 0 & 0 & 0 & \cos\gamma & 0 & 0 & -\sin\gamma \\ 0 & 0 & 0 & 0 & 0 & \cos\gamma & \sin\gamma & 0 \\ 0 & 0 & 0 & 0 & 0 & -\sin\gamma & \cos\gamma & 0 \\ 0 & 0 & 0 & 0 & \sin\gamma & 0 & 0 & \cos\gamma \end{pmatrix} \quad (4.8)$$

$$R_h = \begin{pmatrix} e^{i\varepsilon} & 0 & 0 & 0 & 0 & 0 & 0 & 0 \\ 0 & \cos\varepsilon & 0 & 0 & 0 & \sin\varepsilon & 0 & 0 \\ 0 & 0 & \cos\varepsilon & 0 & 0 & 0 & \sin\varepsilon & 0 \\ 0 & 0 & 0 & \cos\varepsilon & 0 & 0 & 0 & \sin\varepsilon \\ 0 & 0 & 0 & 0 & 1 & 0 & 0 & 0 \\ 0 & -\sin\varepsilon & 0 & 0 & 0 & \cos\varepsilon & 0 & 0 \\ 0 & 0 & -\sin\varepsilon & 0 & 0 & 0 & \cos\varepsilon & 0 \\ 0 & 0 & 0 & -\sin\varepsilon & 0 & 0 & 0 & \cos\varepsilon \end{pmatrix} \quad (4.9)$$

$$R_e = \begin{pmatrix} e^{i\varsigma} & 0 & 0 & 0 & 0 & 0 & 0 & 0 \\ 0 & \cos\varsigma & 0 & 0 & -\sin\varsigma & 0 & 0 & 0 \\ 0 & 0 & \cos\varsigma & 0 & 0 & 0 & 0 & \sin\varsigma \\ 0 & 0 & 0 & \cos\varsigma & 0 & 0 & -\sin\varsigma & 0 \\ 0 & \sin\varsigma & 0 & 0 & \cos\varsigma & 0 & 0 & 0 \\ 0 & 0 & 0 & 0 & 0 & 1 & 0 & 0 \\ 0 & 0 & 0 & \sin\varsigma & 0 & 0 & \cos\varsigma & 0 \\ 0 & 0 & -\sin\varsigma & 0 & 0 & 0 & 0 & \cos\varsigma \end{pmatrix} \quad (4.10)$$

$$R_f = \begin{pmatrix} e^{i\eta} & 0 & 0 & 0 & 0 & 0 & 0 & 0 \\ 0 & \cos\eta & 0 & 0 & 0 & 0 & 0 & -\sin\eta \\ 0 & 0 & \cos\eta & 0 & -\sin\eta & 0 & 0 & 0 \\ 0 & 0 & 0 & \cos\eta & 0 & \sin\eta & 0 & 0 \\ 0 & 0 & \sin\eta & 0 & \cos\eta & 0 & 0 & 0 \\ 0 & 0 & 0 & -\sin\eta & 0 & \cos\eta & 0 & 0 \\ 0 & 0 & 0 & 0 & 0 & 0 & 1 & 0 \\ 0 & \sin\eta & 0 & 0 & 0 & 0 & 0 & \cos\eta \end{pmatrix} \qquad (4.11)$$

$$R_g = \begin{pmatrix} e^{i\theta} & 0 & 0 & 0 & 0 & 0 & 0 & 0 \\ 0 & \cos\theta & 0 & 0 & 0 & 0 & \sin\theta & 0 \\ 0 & 0 & \cos\theta & 0 & 0 & -\sin\theta & 0 & 0 \\ 0 & 0 & 0 & \cos\theta & -\sin\theta & 0 & 0 & 0 \\ 0 & 0 & 0 & \sin\theta & \cos\theta & 0 & 0 & 0 \\ 0 & 0 & \sin\theta & 0 & 0 & \cos\theta & 0 & 0 \\ 0 & -\sin\theta & 0 & 0 & 0 & 0 & \cos\theta & 0 \\ 0 & 0 & 0 & 0 & 0 & 0 & 0 & 1 \end{pmatrix},$$

(4.12)

where h, e , f, and g designate the 5th, 6th, 7th, and 8th dimensions.

Chapter V. Application of □ Matrices in Relativistic Mechanics

Equation Section (Next)

As shown earlier, the Lorentz transformation of Special Relativity can be cast in the form of a four dimensional □ matrix, which will be designated by $\mathcal{L}_l$. (It is noted here that the extension of the Lorentz transform to higher dimensions is straightforward by use of the □ matrix. Extension to higher dimensions, however, involves other complications concerning such things as the proper addition of velocities, for example.) The elements of the $\mathcal{L}_l$ matrix are taken from the four vector of velocity

$$i\gamma c\hat{l} + \gamma\beta_x c\hat{i} + \gamma\beta_y c\hat{j} + \gamma\beta_z c\hat{k} \quad (5.1)$$

This will be the convention in this work, namely the first row of the vector will be the fourth or time component and the subsequent rows the space components in normal xyz order. If each component of the velocity is divided by the magnitude of the 4-velocity, ic , the resulting quantities are the components of the $\mathcal{L}_l$ matrix, thus

$d = i\gamma, a = \gamma\beta_x, b = \gamma\beta_y, c = \gamma\beta_z$ and the Lorentz transform matrix becomes

$$\mathcal{L} = \begin{pmatrix} \gamma & -i\gamma\beta_x & -i\gamma\beta_y & -i\gamma\beta_z \\ i\gamma\beta_x & \gamma & i\gamma\beta_z & -i\gamma\beta_y \\ i\gamma\beta_y & -i\gamma\beta_z & \gamma & i\gamma\beta_x \\ i\gamma\beta_z & i\gamma\beta_y & -i\gamma\beta_x & \gamma \end{pmatrix} \quad (5.2)$$

It seems rather elegant and nice that the Lorentz transform can be constructed so simply from the 4-D velocity vector.

It is easy to see that

$$\mathcal{L}\begin{pmatrix} \text{ict} \\ \text{x} \\ \text{y} \\ \text{z} \end{pmatrix} = \begin{pmatrix} i\gamma\left(ct - \bar{\beta}\Box\bar{r}\right) \\ \gamma\left(x - \beta_x ct + i\beta_z y - i\beta_y z\right) \\ \gamma\left(y - \beta_y ct + i\beta_z x - i\beta_x z\right) \\ \gamma\left(z - \beta_z ct + i\beta_x y - i\beta_y x\right) \end{pmatrix}$$

$$= i\gamma\left(ct - \bar{\beta}\Box\bar{r}\right)\hat{l} + \gamma\left(\bar{r} - \bar{\beta}ct\right) + i\gamma\bar{\beta}\times\bar{r} \qquad (5.3)$$

The last term of this equation is the transformed position 4-vector written in conventional vector notation to show better what has happened. One will notice straightaway that this result appears not to be quite the correct transformed 4-vector and, indeed, further examination produces a very interesting insight. This is the correct result, but not on the usual level. First, we note that the components of the position vector perpendicular to the velocity, $\bar{\beta}$, of the Lorentz transform cannot be affected by the transform, i.e. they must remain unchanged. To proceed, the vector is rewritten in terms of components parallel and perpendicular to the 3-space velocity vector $\bar{\beta}$, thus

$$-i\gamma\left(ct - \beta r_\Box\right)\hat{l} + \gamma\left(r_\Box - \beta ct\right)\hat{p} + \gamma r_\perp \hat{q} - i\gamma\beta r_\perp \hat{s} \qquad (5.4)$$

where the unit vector $\hat{l}$ is in the time direction, the unit vector $\hat{p}$ is parallel to the 3-space velocity, the unit vector $\hat{q}$ is perpendicular to the 3-space velocity , and the unit vector $\hat{s} = \hat{p}\times\hat{q}$ is perpendicular to both $\hat{p}$ and $\hat{q}$. (These symbols are taken from the German parallel, quer, and senkrecht.) Thus we see that all the unit vectors are mutually perpendicular, so that when this vector is squared we get the following:

$$-\gamma^2\left(ct - \beta r_\Box\right)^2 + \gamma^2\left(r_\Box - \beta ct\right)^2 + \gamma^2\left(r_\perp^2 - \beta^2 r_\perp^2\right) \qquad (5.5)$$

Only the perpendicular components, i.e. the last two terms of the 4-vector, can be individually collapsed without effectively collapsing the transformed vector completely to its previous state. Hence, we collapse it partway by collapsing only the last two terms to $r_\perp^2$ giving,

$$-\gamma^2 \left(ct - \beta r_0\right)^2 + \gamma^2 \left(r_0 - \beta ct\right)^2 + r_\perp^2$$

It is clear that this result could also be obtained by squaring the 4-vector below,

$$i\gamma\left(ct - \beta r_0\right)\hat{l} + \gamma\left(r_0 - \beta ct\right)\hat{p} + r_\perp\hat{q}$$

which is the correct Lorentz transform of the time-position 4-vector. But this could also be the vectors

$$i\gamma\left(ct - \beta r_0\right)\hat{l} + \gamma\left(r_0 - \beta ct\right)\hat{p} - r_\perp\hat{q}$$

$$i\gamma\left(ct - \beta r_0\right)\hat{l} - \gamma\left(r_0 - \beta ct\right)\hat{p} + r_\perp\hat{q}$$

$$-i\gamma\left(ct - \beta r_0\right)\hat{l} + \gamma\left(r_0 - \beta ct\right)\hat{p} + r_\perp\hat{q}$$

$$i\gamma\left(ct - \beta r_0\right)\hat{l} - \gamma\left(r_0 - \beta ct\right)\hat{p} - r_\perp\hat{q}$$

$$-i\gamma\left(ct - \beta r_0\right)\hat{l} - \gamma\left(r_0 - \beta ct\right)\hat{p} + r_\perp\hat{q}$$

etc.

(This is a bit reminiscent of quantum mechanics.) We are lacking a mathematical mechanism to weed out the spurious vectors, instead we must think. If we let $\beta \to 0$, then the signs of the resulting components must be the same as before the transform, right? Thus in the above example we started with the vector $ict\hat{l} + \bar{r}$ so we must get the same vector. This rules out inversions of the time and leaves the possibilities

$$i\gamma\left(ct - \beta r_0\right)\hat{l} + \gamma\left(r_0 - \beta ct\right)\hat{p} + r_\perp\hat{q}$$

$$i\gamma\left(ct - \beta r_0\right)\hat{l} + \gamma\left(r_0 - \beta ct\right)\hat{p} - r_\perp\hat{q}$$

$$i\gamma\left(ct - \beta r_0\right)\hat{l} - \gamma\left(r_0 - \beta ct\right)\hat{p} + r_\perp\hat{q}$$

$$i\gamma\left(ct - \beta r_0\right)\hat{l} - \gamma\left(r_0 - \beta ct\right)\hat{p} - r_\perp\hat{q}$$

However, if we take $\bar{r}$ to be a radius vector from some origin, then its components must be positive before and after the transform, consequently we are left only the possibility of $i\gamma\left(ct - \beta r_0\right)\hat{l} + \gamma\left(r_0 - \beta ct\right)\hat{p} + r_\perp\hat{q}$, which is the correct vector. So all is okay for the time-position 4-vector.

This complicates the transform as one must sometimes unsnarl terms and square and unsquare the transformed vector. Nevertheless with a little care, the correct components can be determined.

The Lorentz transform, equation (5.2), can be rewritten in a pseudo two dimensional form, thus

$$\mathcal{L}_1 = \begin{pmatrix} \gamma & -i\gamma\bar{\beta} \\ i\gamma\bar{\beta} & \gamma \end{pmatrix} \qquad (5.6)$$

So that

$$\mathcal{L}_1 \begin{pmatrix} ict \\ x \\ y \\ z \end{pmatrix} = \begin{pmatrix} \gamma & -i\gamma\bar{\beta} \\ i\gamma\bar{\beta} & \gamma \end{pmatrix} \begin{pmatrix} ict \\ \bar{r} \end{pmatrix} = \begin{pmatrix} i\gamma\left(ct - \bar{\beta}\Box\bar{r}\right) \\ -\gamma\bar{\beta}ct + \gamma\bar{r} - i\gamma\bar{\beta}\times\bar{r} \end{pmatrix}$$

Which is the same result as equation (5.3). This pseudo two dimensional equation can, in turn be written in terms of the vectors $\hat{I}$, $\hat{p}$, $\hat{q}$, and $\hat{r}$ defined above:

$$\begin{pmatrix} \gamma\hat{I} & -i\gamma\beta\hat{p} \\ i\gamma\beta\hat{p} & \gamma\hat{I} \end{pmatrix} \begin{pmatrix} ict\hat{I} \\ r_\Box\hat{p} + r_\perp\hat{q} \end{pmatrix} = \begin{pmatrix} \left(i\gamma ct - i\gamma\beta r_\Box\hat{p}\Box\hat{p} - i\gamma\beta r_\perp\hat{p}\Box\hat{q}\right)\hat{I} \\ -\gamma\beta ct\hat{p} + \gamma\left(r_\Box\hat{p} + r_\perp\hat{q}\right) - i\gamma\beta r_\perp\hat{s} \end{pmatrix}$$

$$= \begin{pmatrix} i\gamma\left(ct - \beta r_\Box\right)\hat{I} \\ \gamma\left(r_\Box - \beta ct\right)\hat{p} + \gamma r_\perp\hat{q} - i\gamma\beta r_\perp\hat{s} \end{pmatrix} = \begin{pmatrix} i\gamma\left(ct - \beta r_\Box\right)\hat{I} \\ \gamma\left(r_\Box - \beta ct\right)\hat{p} + \gamma r_\perp\left(\hat{q} - i\beta\hat{s}\right) \end{pmatrix}$$

(5.7)

Where here we take $\hat{I}$ times any vector to be that vector, i.e. $\hat{I}$ multiplies like an identity vector. This technique gives us straightaway the equation (5.4). It is usually much easier to see how the Lorentz transformed 4-vector unsnarls when the transform is done in this way, i.e. any vector terms that can be factored to a scalar times a vector, such as $\gamma(\hat{q} - i\beta\hat{s})$ or such as $\gamma\left(i\beta\hat{I} - \hat{p}\right)$, can be reduced or collapsed to the scalar times $\hat{q}$ or respectively the scalar times $\hat{p}$. It is easy to become adept at this technique of operating $\Box$ matrices on 4-vectors; consequently the author will use this technique quite frequently in the rest of this book; because of its convenience and rapidity.

27

Note that the Lorentz transformed space vector satisfies the additional condition that

$$\left(ict'\right)^2 + \bar{r}'^2 = \left(ict\right)^2 + \bar{r}^2. \qquad (5.8)$$

Equation (5.8) is an important condition that must in some way be met by the 4-vectors of relativistic physics. Also, note that, in the case of the Lorentz transformation, the components of the position 4-vectors, both transformed and untransformed, set equal to constants are, of course, characteristics of the wave equation.

The result that was obtained above implies that the time direction is somehow related or parallel to the 3-velocity direction because the parallel component of the position vector is included in the time component. What that may mean, if anything, is not yet clear, however.

If

$$ct' = \gamma\left(ct - \bar{\beta}\Box\bar{r}\right)$$
$$\bar{r}' = \gamma\left(\bar{r} - \bar{\beta}ct\right) - i\gamma\left(\bar{\beta}\times\bar{r}\right) \qquad (5.9),$$

then it is easy to show that

$$ct = \gamma\left(ct' + \bar{\beta}\Box\bar{r}'\right)$$
$$\bar{r} = \gamma\left(\bar{r}' + \bar{\beta}ct'\right) + i\gamma\left(\bar{\beta}\times\bar{r}'\right) \qquad (5.10).$$

That is, the transform on primed vectors with velocity reversed is reasonably symmetrical.

The next example we take is the electromagnetic field 4-vector. From my theory this is

$$iL\hat{I} + \bar{E} + i\bar{B} \quad where \quad L = \frac{1}{c}\frac{\partial\phi}{\partial t} + \nabla\Box\bar{A}.$$

I call L the Lorentz factor. I make it a rule that imaginary time components plus real spatial components make a 4-vector that is distinct from a 4- vector that has a real time component and imaginary space components. This is similar to Hermitian and Anti Hermitian matrices. Now because of gauge invariance we can impose the Lorentz condition, namely $L = 0$, so we only have to transform the three vector $\bar{E} + i\bar{B}$, thus

28

$$\mathcal{L}\left(\overline{E}+i\overline{B}\right)=\gamma\left(-i\overline{\beta}\square\overline{E}+\overline{\beta}\square\overline{B}\right)\hat{I}+\gamma\overline{E}+i\gamma\overline{B}-i\gamma\overline{\beta}\times E+\gamma\overline{\beta}\times\overline{B}$$

Collecting components into the two types of 4-vector, one gets,

$$-i\gamma\left(\overline{\beta}\square\overline{E}\right)\hat{I}+\gamma\left(\overline{E}+\overline{\beta}\times\overline{B}\right)+\gamma\left(\overline{\beta}\square\overline{B}\right)\hat{I}+i\gamma\left(\overline{B}-\overline{\beta}\times\overline{E}\right)$$

As before, we sort out the components parallel and perpendicular to the velocity, thus

$$-i\gamma\left(\beta E_{\square}\right)\hat{I}+\gamma E_{\perp}\hat{q}+\gamma\left(\overline{\beta}\times\overline{B}\right)+\gamma E_{\square}\hat{p}+\gamma\left(\beta B_{\square}\right)\hat{I}+i\gamma B_{\perp}\hat{q}-i\gamma\left(\overline{\beta}\times\overline{E}\right)+i\gamma B_{\square}\hat{p}$$

The E-M field cannot have a time component, so we must collapse the parallel components, thus squaring,

$$-\gamma^{2}\left(\beta E_{\square}\right)^{2}+\gamma^{2}\left(E_{\perp}\hat{q}+\left(\overline{\beta}\times\overline{B}\right)\right)^{2}+\gamma^{2}E_{\square}^{2}+\gamma^{2}\left(\beta B_{\square}\right)^{2}-\gamma^{2}\left(B_{\perp}\hat{q}-\left(\overline{\beta}\times\overline{E}\right)\right)^{2}-\gamma^{2}B_{\square}^{2}$$

And collapsing,

$$E_{\square}^{2}+\gamma^{2}\left(E_{\perp}\hat{q}+\left(\overline{\beta}\times\overline{B}\right)\right)^{2}-B_{\square}^{2}-\gamma^{2}\left(B_{\perp}\hat{q}-\left(\overline{\beta}\times\overline{E}\right)\right)^{2}$$

Unsquaring yields

$$E_{\square}\hat{p}+\gamma\left(E_{\perp}\hat{q}+\overline{\beta}\times\overline{B}\right)+iB_{\square}\hat{p}+i\gamma\left(B_{\perp}\hat{q}-\overline{\beta}\times\overline{E}\right)$$

where we have chosen the signs of the parallel and perpendicular components to correspond to those of the untransformed parallel and perpendicular components. These look a lot like the equations in Jackson.

The Galilean transform of classical mechanics can be recovered from equation (5.2) by allowing c, the velocity of light to approach infinity in such a way that $\gamma\rightarrow 1$ (this is equivalent to making the retardation time, $t_{r}=\dfrac{r}{c}$, negligible) and

$$\gamma\begin{pmatrix} i & \beta_{x} & \beta_{y} & \beta_{z} \\ -\beta_{x} & i & -\beta_{z} & \beta_{y} \\ -\beta_{y} & \beta_{z} & i & -\beta_{x} \\ -\beta_{z} & -\beta_{y} & \beta_{x} & i \end{pmatrix}\rightarrow$$

$$-\frac{i}{c}\begin{pmatrix} ic & \overline{v}_x & \overline{v}_y & \overline{v}_z \\ -\overline{v}_x & ic & -\overline{v}_z & \overline{v}_y \\ -\overline{v}_y & \overline{v}_z & ic & -\overline{v}_x \\ -\overline{v}_z & -\overline{v}_y & \overline{v}_x & ic \end{pmatrix}\begin{pmatrix} ict \\ x \\ y \\ z \end{pmatrix} = \frac{1}{c}\left[i\left(c^2t - \overline{v}\cdot\overline{r}\right)\hat{i} + \left(c\overline{r} - \overline{v}ct\right) + \overline{v}\times\overline{r}\right] =$$

$$i\left(ct - \frac{\overline{v}\cdot\overline{r}}{c}\right)\hat{i} + \left(\overline{r} - \overline{v}t\right) + \frac{\overline{v}}{c}\times\overline{r} \rightarrow ict\hat{i} + \overline{r} - \overline{v}t = ict\hat{i} + \left(r_{\square} - vt\right)\hat{p} + r_{\perp}\hat{q}$$

(5.11)

where c is very large compared to 1. The space component of equation (5.11) is the correct unsnarled Galilean result.

A brief digression is made here to compare equation (5.2) above to the conventional "generalized" Lorentz transform as given by equation (11.21) on page 357 of Jackson (Reference 4 - John David Jackson, *Classical Electrodynamics,* John Wiley and Sons, New York (1966)). First we rewrite Jackson's equation in terms of $\overline{r}$ and the reduced velocity, $\overline{\beta}$, instead of $\overline{x}$ and velocity, so that

$$\overline{r}' = \overline{r} + (\gamma - 1)\frac{\overline{\beta}\cdot\overline{r}}{\beta^2}\overline{\beta} - \gamma\overline{\beta}ct$$

$$ct' = \gamma\left(ct - \overline{\beta}\cdot\overline{r}\right)$$

The time component is essentially the same as our result; so we have only to investigate whether $\overline{r}'$ is basically the same or not. One should note that squaring, unsnarling, and unsquaring both the author's equation and Jackson's equation gives the same result. Rearranging terms in Jackson's equation for $\overline{r}'$, we get

$$\overline{r}' = \gamma\left(\overline{r} - \overline{\beta}ct\right) + (\gamma - 1)\left(\frac{\overline{r}\cdot\overline{\beta}}{\beta^2}\overline{\beta} - \overline{r}\right)$$

$$\overline{r}' = \gamma\left(\overline{r} - \overline{\beta}ct\right) + \frac{(\gamma - 1)}{\beta^2}\left(\overline{\beta}\times(\overline{\beta}\times\overline{r})\right)$$

This compares to the author's result,

$$\overline{r}' = \gamma\left(\overline{r} - \overline{\beta}ct\right) + i\gamma\left(\overline{\beta}\times\overline{r}\right).$$

Clearly Jackson's result is more convoluted.

To be completely fair, it is shown that squaring, unsnarling and unsquaring Jackson's equation gives the correct result. As pointed out above it is necessary to deal with only the expression for $\vec{r}'$:

$$\vec{r}' = \frac{\beta^2 \vec{r} - (\vec{\beta} \cdot \vec{r})\vec{\beta}}{\beta^2} + \gamma \left(\frac{\vec{\beta} \cdot \vec{r}}{\beta^2} \vec{\beta} - \vec{\beta} ct \right)$$

$$= -\frac{\vec{\beta} \times (\vec{\beta} \times \vec{r})}{\beta^2} + \frac{\vec{\beta}}{\beta^2} \left[\vec{\beta} \left(\gamma \left(\vec{r} - \vec{\beta} ct \right) \right) \right]$$

(5.12).

The second term on the right is the component parallel to the velocity, the same as what would be obtained from equation (5.2). It remains to show that the first term on the right is the component perpendicular to the velocity. Squaring the first term and using a little algebraic manipulation, we get

$$\frac{\gamma^2 (1 - \beta^2)(\vec{\beta} \times (\vec{\beta} \times \vec{r}))^2}{\beta^4} = \frac{\gamma^2 (\vec{\beta} \times (\vec{\beta} \times \vec{r}))^2}{\beta^4} - \frac{\gamma^2 (\vec{\beta} \times (\vec{\beta} \times \vec{r}))^2}{\beta^2}$$

(5.13)

Unsquaring the two terms above separately, we get finally the perpendicular components

$$\left(\frac{\vec{\beta} \times (\vec{\beta} \times)}{\beta^2} \right) \gamma(\vec{r}) - \left(\frac{\vec{\beta} \times}{\beta} \right) \gamma (\vec{\beta} \times \vec{r}) \quad (5.14)$$

These are essentially the same as from equation (5.3). We return now to the main discussion.

A second Lorentz transform is made on the previously Lorentz transformed position 4-vector to see if the correct velocity addition rule is obtained, thus

$$-i\begin{pmatrix} i\gamma_\alpha & \gamma_\alpha\alpha_x & \gamma_\alpha\alpha_y & \gamma_\alpha\alpha_z \\ -\gamma_\alpha\alpha_x & i\gamma_\alpha & -\gamma_\alpha\alpha_z & \gamma_\alpha\alpha_y \\ -\gamma_\alpha\alpha_y & \gamma_\alpha\alpha_z & i\gamma_\alpha & -\gamma_\alpha\alpha_x \\ -\gamma_\alpha\alpha_z & -\gamma_\alpha\alpha_y & \gamma_\alpha\alpha_x & i\gamma_\alpha \end{pmatrix}\begin{pmatrix} i\gamma_\beta\left(ct - \overline{\beta}\cdot\overline{r}\right) \\ \gamma_\beta(x - \beta_x ct) - i\gamma_\beta(\beta_y z - \beta_z y) \\ \gamma_\beta(y - \beta_y ct) - i\gamma_\beta(\beta_z x - \beta_x z) \\ \gamma_\beta(z - \beta_z ct) - i\gamma_\beta(\beta_x y - \beta_y x) \end{pmatrix} =$$

$$i\gamma_\alpha\gamma_\beta\left[\overline{r}\left(1 + \overline{\alpha}\cdot\overline{\beta}\right) - \left(\overline{\alpha} + \overline{\beta} + i\overline{\alpha}\times\overline{\beta}\right)\cdot\overline{r}\right]\hat{i}$$
$$+\gamma_\alpha\gamma_\beta\left[ct\left(1 + \overline{\alpha}\cdot\overline{\beta}\right) - \left(\overline{\alpha} + \overline{\beta} + i\overline{\alpha}\times\overline{\beta}\right)ct\right]$$
$$-i\gamma_\alpha\gamma_\beta\left(\overline{\alpha} + \overline{\beta} + i\overline{\alpha}\times\overline{\beta}\right)\times\overline{r}$$

$$(5.15)$$

Clearly, the sum of the velocities $\overline{\alpha}$ and $\overline{\beta}$ is

$$\overline{u} = \frac{\overline{\alpha} + \overline{\beta} + i(\overline{\alpha}\times\overline{\beta})}{1 + \overline{\alpha}\cdot\overline{\beta}} \qquad (5.16)$$

and

$$\gamma_u = \gamma_\alpha\gamma_\beta(1 + \overline{\alpha}\cdot\overline{\beta}). \qquad (5.17)$$

Equations (5.16) and (5.17) will hereafter be called the Special Relativity Velocity Addition Rule, SRVAR for short. The presence of the cross product in equation (5.16) makes the sum $\overline{u}$ different depending on the order of addition of $\overline{\alpha}$ and $\overline{\beta}$. This is alright as long as the magnitude of the sum, u, is the same regardless of order of addition. If $\overline{\alpha}$ and $\overline{\beta}$ are parallel, then we get the conventional result, namely

$$\overline{u} = \frac{\alpha + \beta}{1 + \alpha\beta}\hat{p} \qquad (5.18)$$

If $\overline{\alpha}$ and $\overline{\beta}$ are perpendicular, then

$$\overline{u} = \alpha\hat{p} + \beta\hat{q} + i\alpha\beta\hat{s} \qquad (5.19)$$

where $\hat{s} = \hat{p}\times\hat{q}$ as before, so that squaring and unsquaring $\overline{u}$, i.e. finding its magnitude, we get

$$u = \sqrt{\alpha^2 + \beta^2 - \alpha^2\beta^2} \qquad (5.20)$$

If $\overline{\alpha}$ is at an angle θ with $\overline{\beta}$, then

$$u = \sqrt{1 + \frac{(\alpha^2 - 1)(1 - \beta^2)}{(1 + \alpha\beta\cos\theta)^2}} \qquad (5.21)$$

and so on. Note that in every case, if either α or β or both α and β are equal to 1, then u is equal to 1, as expected. Note also that, if either α or β are equal to 1, addition of any negative velocity to it does not change the magnitude of u which remains equal to 1. This implies that anything moving at the velocity of light in vacuum, c, cannot be accelerated (deceleration, or also acceleration from a velocity slightly less than c, is another matter). But that is not all! Equation (5.16) shows that it also does not matter what the directions of $\overline{\alpha}$ and $\overline{\beta}$ might be, the magnitude of $\overline{u}$ cannot be greater than one. The $i\overline{\alpha} \times \overline{\beta}$ term is therefore essential, but missing from the current formulas for the relativistic sum of velocities!

Addition of more than two velocities, leads to conditions on various components of the velocities. Also, it is clear that conventional vector addition of velocity components is, at least to some extent, inconsistent with the velocity addition rule of Special Relativity. These considerations are taken up in Chapter IX *Further Development of Ideas – The Special Relativity Velocity Addition Rule*.

One should carefully note that use of a matrix such as

$$\begin{pmatrix} d & a & b & c \\ a & -d & -c & b \\ b & c & -d & -a \\ c & -b & a & -d \end{pmatrix} \text{ or}$$

$$\begin{pmatrix} d & a & b & c \\ a & -d & c & -b \\ b & -c & -d & a \\ c & b & -a & -d \end{pmatrix}, \text{ does not give the}$$

correct velocity addition rule, even though these matrices yield the correct E-M fields and Maxwell equations when applied to a proper potential 4-vector.

Now consider an uncharged elementary particle of rest mass = m_0. If it is assumed that such a nonzero rest mass particle has, in the rest frame, a "rest momentum" in the time direction given by $im_0 c\hat{l}$ and our Lorentz transform matrix $\mathcal{L}_l$ is applied to this rest momentum, thus

$$\mathcal{L}_l \qquad\qquad (im_0 c\hat{l}) \qquad\qquad =$$

$$-i\begin{pmatrix} i\gamma & \gamma\beta_x & \gamma\beta_y & \gamma\beta_z \\ -\gamma\beta_x & i\gamma & -\gamma\beta_z & \gamma\beta_y \\ -\gamma\beta_y & \gamma\beta_z & i\gamma & -\gamma\beta_x \\ -\gamma\beta_z & -\gamma\beta_y & \gamma\beta_x & i\gamma \end{pmatrix}\begin{pmatrix} im_0 c \\ 0 \\ 0 \\ 0 \end{pmatrix} = i\gamma m_0 c\hat{l} + i\gamma mc_0 \overline{\beta} \text{ ,(5.22)}$$

the correct 4-momentum vector of special relativity is recovered. This result lends credence to the existence of a rest momentum for any elementary particle of finite rest mass. If we multiply the energy-momentum vector by c, the velocity of light, and square the resulting vector, we get

$$-\gamma^2 m_0^2 c^4 + \gamma^2 m_0^2 c^4 \beta^2 = -m_0^2 c^4 . \qquad (5.23)$$

If we designate the total energy of the particle by $\mathcal{E}$, then

$$\mathcal{E}^2 = c^2 p^2 + m_0^2 c^4 \qquad\qquad (5.24)$$

the normal energy-momentum relation of special relativity is thereby obtained. The momentum 4-vector is

$^4\overline{p} = i\dfrac{\mathcal{E}}{c}\hat{l} + \overline{p}$ so that, if we call the 4-momentum $^4\overline{p}$, where

$\left|^4\overline{p}\right| = m_0 c$, then we can write for the 4-momentum

$$im_0 c\hat{\sigma} = i\dfrac{\mathcal{E}}{c}\hat{l} + \overline{p} \qquad (5.25)$$

Application of i times the D_l^* matrix with the time element, d, changed to the total instead of the partial time derivative, to the energy vector yields

$$D\left(i\mathcal{E}\hat{l} + c\overline{p}\right) = i\left(\dfrac{1}{c}\dfrac{\partial\mathcal{E}}{\partial t} + c\nabla\square\overline{p}\right)\hat{l} + \left(\nabla\mathcal{E} + \dfrac{d\overline{p}}{dt}\right) + ic\nabla\times\overline{p} . \ (5.26)$$

34

This equation is the classical relativistic force equations collected together in one equation in 4-vector form, including a momentum vorticity term.

If one constructs a special □ matrix D operator from derivatives with respect to the generalized coordinates of Hamilton and of time, namely:

$$D_q = \begin{pmatrix} \dfrac{\partial}{\partial t} & -i\nabla_q \\ i\nabla_q & \dfrac{\partial}{\partial t} \end{pmatrix}, \; then$$

$$D_q \begin{pmatrix} i\mathcal{E}\hat{i} \\ \overline{p} \end{pmatrix} = \begin{pmatrix} i\left(\dfrac{\partial \mathcal{E}}{\partial t} - \nabla_q \Box \overline{p}\right) & \hat{i} \\ \nabla_q \mathcal{E} + \dfrac{\partial \overline{p}}{\partial t} - ic\nabla_q \times \overline{p} \end{pmatrix} \quad (5.27)$$

Provided $\nabla_q \mathcal{E} + \dfrac{\partial \overline{p}}{\partial t} = 0$, the total energy, $\mathcal{E}$, is replaced by the Hamiltonian function, H, and $\overline{r}$ and $\overline{p}$ by the canonical coordinates q_i and by the canonical momenta p_i, one gets

$$\frac{\partial \mathcal{E}}{\partial q_i} = \frac{\partial p_i}{\partial t}$$

This is one of Hamilton's canonical equations. Also if one constructs a special □ matrix D operator from derivatives with respect to the generalized momenta, namely

$$D_p = \begin{pmatrix} \dfrac{\partial}{\partial p_4} & -i\nabla_p \\ i\nabla_p & \dfrac{\partial}{\partial p_4} \end{pmatrix}, \; then$$

$$D_p \begin{pmatrix} i\mathcal{E}\hat{i} \\ c\overline{p} \end{pmatrix} = \begin{pmatrix} i\left(\dfrac{\partial \mathcal{E}}{\partial p_4} - c\nabla_p \Box \overline{p}\right) & \hat{i} \\ -\nabla_p \mathcal{E} + c\dfrac{\partial \overline{p}}{\partial p_4} - ic\nabla_p \times \overline{p} \end{pmatrix} \quad (5.28)$$

Provided that $-\nabla_p \mathcal{E} + c \dfrac{\partial \overline{p}}{\partial p_4} = 0$, one gets

$$\frac{\partial \mathcal{E}}{\partial p_i} = c \frac{\partial \overline{p}}{\partial p_4}$$

But $p_4 = mc$ and c has to be a constant, hence $\dfrac{\partial}{\partial p_4}$ could only

be equal to $\dfrac{1}{c} \dfrac{\partial}{\partial m}$ and

$$\frac{\partial \mathcal{E}}{\partial p_i} = \frac{\partial p_i}{\partial m}$$

However, $p_i = m \dfrac{\partial q_i}{\partial t}$, so that, because $q_i \neq q_i(m)$, $\dfrac{\partial p_i}{\partial m} = \dfrac{\partial q_i}{\partial t}$
and consequently,

$$\frac{\partial \mathcal{E}}{\partial p_i} = \frac{\partial q_i}{\partial t} \qquad (5.29)$$

This, of course, is another one of Hamilton's canonical equations.

Finally, note also that in the limit that the velocity is much less than c, we can set $\mathcal{E} = \gamma m_0 c^2 - V$, where V is the potential energy, or

$\mathcal{E} = \gamma m_0 c^2 - e\phi$ and $\overline{p} \Rightarrow \overline{p} - \dfrac{e}{c}\overline{A}$, where $e\phi$ and $e\overline{A}$ are the

potentials due to an E-M field (see Reference 5 - Leon Brillouin, *Relativity Reexamined,* Academic Press, New York, (1970), Chapter 2, p. 13 - 28).

Chapter VI. Application of ☐ Matrices in Electromagnetic Theory

Equation Section (Next)

I begin with the 4-vector potential energy obtained from the potential of a static single charge distribution, $e\phi$. Now the "static" potential, or at least the charge e, is unchanged by its velocity; therefore, strangely I might add, the 4-vector potential energy can be obtained (this is unconventional) by use of an "infinitesimal(?)", or perhaps better, a "charge" Lorentz transformation wherein $\overline{\beta}$ is not small, but $\gamma = 1$. Hence, one gets

$$\mathcal{L}_v = -i \begin{pmatrix} i & \beta_x & \beta_y & \beta_z \\ -\beta_x & i & -\beta_z & \beta_y \\ -\beta_y & \beta_z & i & -\beta_x \\ -\beta_z & -\beta_y & \beta_x & i \end{pmatrix} \quad (6.1) \text{ and}$$

$$\mathcal{L}_v \left(ie\phi\hat{l} \right) = ie\phi\hat{l} + e\phi\overline{\beta}.$$

This can be generalized, of course, by summing over all elementary charges having different velocities to obtain the electromagnetic potential energy 4-vector, $i\phi\hat{l} + \overline{A}$, (absorbing the charge into ϕ and $\overline{A}$) as seen by a stationary observer. This external potential energy 4-vector can be operated on with the conjugate of the D_1 matrix to get the E-M fields external to the observer, thus

$$D_1^*(i\phi\hat{l} + \overline{A}) = \begin{pmatrix} -\dfrac{i}{c}\dfrac{\partial}{\partial t} & \dfrac{\partial}{\partial x} & \dfrac{\partial}{\partial y} & \dfrac{\partial}{\partial z} \\[2mm] -\dfrac{\partial}{\partial x} & -\dfrac{i}{c}\dfrac{\partial}{\partial t} & \dfrac{\partial}{\partial z} & \dfrac{\partial}{\partial y} \\[2mm] -\dfrac{\partial}{\partial y} & \dfrac{\partial}{\partial z} & -\dfrac{i}{c}\dfrac{\partial}{\partial t} & -\dfrac{\partial}{\partial x} \\[2mm] -\dfrac{\partial}{\partial z} & -\dfrac{\partial}{\partial y} & \dfrac{\partial}{\partial x} & -\dfrac{i}{c}\dfrac{\partial}{\partial t} \end{pmatrix} \begin{pmatrix} i\phi \\ A_X \\ A_Y \\ A_Z \end{pmatrix} =$$

$$\left(\frac{1}{c}\frac{\partial \phi}{\partial t} + \nabla \Box \overline{A}\right)\hat{l} - i(\nabla \phi + \frac{1}{c}\frac{\partial \overline{A}}{\partial t}) + \nabla \times \overline{A} = L\hat{l} + i\overline{E} + \overline{B}, \ (6.2)$$

where $\overline{E}$ and $\overline{B}$ are the electric and magnetic field vectors respectively and $L\hat{l}$ is the time component, $L = \dfrac{1}{c}\dfrac{\partial \phi}{\partial t} + \nabla \Box \overline{A}$, which I will call the Lorentz factor. We know by previous trial that we must first use the conjugate operator D_1^* to get the correct results. In fact, repeated applications of the differential operator require alternation between the conjugate operators D_1^* and D_1 to get the correct differential equations. In order to set the stage for an interesting insight into the E-M fields, it is necessary to discuss also the Lorentz condition, which is related to the gauge invariance of the E-M fields. By the theorem, that a vector is uniquely specified by its divergence and curl within a region and its normal component over the boundary, we can choose $\nabla \Box \overline{A}$ with considerable latitude without loss of generality, i.e. the measurable E-M fields, $\overline{E}$ and $\overline{B}$ will remain the same. For the moment I choose $\nabla \Box \overline{A}$ such that the Lorentz factor, L, is not zero. At this point, one of the peculiarities I mentioned in Chapter III has cropped up: namely, the new E-M field vector that I have obtained consists of two vectors. The first vector is $L\hat{l} + i\overline{E}$ and the second vector is $\overline{B}$ (which also means that the magnetic field has no time or 4th component (an important result that will be discussed a little later in the section)). This dichotomy appears to arise from the cross product side of the $\Box$ matrix, because the difference between operating with D_1^* or D_1 is that one produces fields that differ one from the other only by a space inversion (a time inversion will not do). Thus

$$D_1(i\phi\hat{l} + \overline{A}) = \left(-\frac{1}{c}\frac{\partial\phi}{\partial t} + \nabla\overline{\Box A}\right)\hat{l} - i(\nabla\phi - \frac{1}{c}\frac{\partial\overline{A}}{\partial t}) + \nabla\times\overline{A} \qquad (6.3)$$

From which we see that inversion of the space coordinates gives

$$\left(-\frac{1}{c}\frac{\partial\phi}{\partial t} - \nabla\overline{\Box A}\right)\hat{l} + i(\nabla\phi + \frac{1}{c}\frac{\partial\overline{A}}{\partial t}) - \nabla\times\overline{A} = -L\hat{l} - i\overline{E} - \overline{B}, \quad (6.4)$$

which are the negatives of the fields obtained by operation with D_1^*.

Returning to the repeated operation with the differential operator, we see that

$$D_1 D_1^*(i\phi\hat{l} + \overline{A}) = D_1(L\hat{l} + i\overline{E} + \overline{B}) =$$

$$[i(\frac{1}{c}\frac{\partial L}{\partial t} + \nabla\overline{\Box E}) + \nabla\overline{\Box B}]\hat{l} - \nabla L - \frac{1}{c}\frac{\partial\overline{E}}{\partial t} + \frac{i}{c}\frac{\partial\overline{B}}{\partial t} + i\nabla\times\overline{E} + \nabla\times\overline{B} \quad .(6.5)$$

and, if this vector is set equal to the charge-current density 4-vector, $i4\pi\rho\hat{l} + \dfrac{4\pi}{c}\overline{j}$, the Maxwell equations are obtained. Note that the non-zero Lorentz factor contributes a charge density $\rho_e = \dfrac{1}{4\pi c}\dfrac{\partial L}{\partial t}$ and current density $\overline{j}_e = -\dfrac{c\nabla L}{4\pi}$ extraneous (at least for E-M fields) to the normal charge and current densities ρ and $\overline{j}$. It is concluded, therefore, that the Lorentz condition, $L = 0$, must apply. But this means that the 4th component of the electric field must be zero, and taken with the previous observation that the magnetic field has no 4th component, we find that the electromagnetic field has no capability of producing a 4th component of force, i.e. there can be no E-M interaction in the direction of time. Also it appears that the electric and magnetic fields should be dealt with together and not separately. If only an electric field is present, then $\overline{A} = 0$ and equation (5.2) becomes $D_1\left(i\phi\hat{l}\right) = -\dfrac{1}{c}\dfrac{\partial\phi}{\partial t}\hat{l} - i\nabla\phi$. But an electric field existing all by itself can only be a static field; and therefore $\dfrac{\partial\phi}{\partial t} = 0$; that is, the fourth component of the field is again zero. On the other hand, if only a magnetic field is present, then $\phi = 0$ and $\dfrac{\partial A}{\partial t} = 0$, so that $D_1\left(\overline{A}\right) = \left(\nabla \square \overline{A}\right)\hat{i} + \nabla \times \overline{A}$.

However, we can still choose $\nabla \square \overline{A} = 0$ and again the fourth component of the field is zero. This result is remarkable, but is consistent with the rest momentum of a non zero rest mass particle having a 4th component of momentum that is equal to $m_0 c$. A rest momentum of $m_0 c$ implies that a non-zero rest mass particle cannot be accelerated in the time direction; however, its mass can be changed. Ordinary creatures on this planet can directly sense only E-M and gravitational forces (also gauge invariant), as far as we now know. Consequently, with the lack of any direct interactions in the time direction, one could not expect them, man included, to have evolved any direct sense of a fourth dimension. We cannot tell the direction of time; we can only tell the passage of time (and that by change only).

The question of whether the E-M field has a fourth component, needless to say, requires further discussion. If there is no fourth component of the E-M field under any, or even normal, circumstances, then it must not be possible to transform the E-M field in any way that produces a fourth component. The larger question of whether, in general, a fourth component of any of the known forces (E-M, Weak, Strong, or Gravitational (which are all, by the way, gauge invariant, and therefore the author suspects that they also have zero fourth components)) exists or whether there can be any kind of force in the time direction is left open at this time (see Chapter XII, *"Recapitulation"*).

Continuing with successive operations of the D_1 and the D_1^* matrices, we further get

$$D_1^* D_1 D_1^* (i\phi\hat{i} + \overline{A}) = D_1^*(i4\pi\rho\hat{i} + \frac{4\pi \overline{j}}{c}) = \frac{4\pi}{c}(\frac{\partial \rho}{\partial t} + \nabla \square \overline{j})\hat{i} - i4\pi(\nabla \rho + \frac{1}{c^2}\frac{\partial \overline{j}}{\partial t}) + \frac{4\pi}{c}\nabla \times \overline{j} \quad (6.6)$$

which is a general charge-current density conservation law. It is fairly easy to show that

$$4\pi\left(\nabla \rho + \frac{1}{c^2}\frac{\partial \overline{j}}{\partial t}\right) = \nabla^2 \overline{E} - \frac{1}{c^2}\frac{\partial^2 \overline{E}}{\partial t^2} \quad and$$

$$\frac{4\pi}{c}\nabla\times\overline{j}=-\nabla^2\overline{B}+\frac{1}{c^2}\frac{\partial^2\overline{B}}{\partial t^2}$$ and, of course, the charge-current continuity equation

$$\frac{\partial\rho}{\partial t}+\nabla\square\overline{j}=0,$$ if no sources or sinks are present.

Operation of D_1 on the above equation carries us to a reasonable limit on repeated operations of the D_1 matrices, namely,

$$D_1 D_1^* D_1 D_1^* (i\phi\hat{l}+\overline{A})=D_1[\frac{4\pi}{c}(\frac{\partial\rho}{\partial t}+\nabla\square\overline{j})\hat{l}-4\pi(\nabla\rho+\frac{1}{c^2}\frac{\partial\overline{j}}{\partial t})+\frac{4\pi}{c}\nabla\times\overline{j}]$$

$$=4\pi i(\frac{1}{c^2}\frac{\partial^2\rho}{\partial t^2}-\nabla^2\rho)\hat{l}+\frac{4\pi}{c}(-\nabla^2\overline{j}+\frac{1}{c^2}\frac{\partial^2\overline{j}}{\partial t^2})$$

(6.7).

Now let us apply the Lorentz transform operator to the E-M field. To get the correct result if we transform with the $\mathcal{L}_1$ matrix, the E-M field vector must be written in the form $iL\hat{l}+\overline{E}+i\overline{B}$. Hence,

$$\mathcal{L}_1(iL\hat{l}+\overline{E}+i\overline{B})=[-\gamma(L-\overline{\beta}\square\overline{E})+i\gamma(\overline{\beta}\square\overline{B})]\hat{l}+i\gamma(\overline{E}-L\overline{\beta}+\overline{\beta}\times\overline{B})-\gamma(\overline{B}-\overline{\beta}\times\overline{E})]$$ (6.8)

Note that $$\left[\mathcal{L}_1\begin{bmatrix}iL\hat{l}\\\overline{E}+i\overline{B}\end{bmatrix}\right]^2=\gamma^2\left[-L^2(1-\beta^2)=E^2(1-\beta^2)-B^2(1-\beta^2)\right]=-L^2+E^2-B^2.$$

Once again we see that the presence of a non-zero Lorentz factor gives rise to extraneous terms in the electric field. To get the correct result, L must be set equal to zero. Now the Electromagnetic field propagates at the velocity of light; hence, adding a velocity $\overline{\beta}\neq 0$ to it does not change its propagation velocity. On this basis (and also the principle that there is no fourth component of the E-M field), it is expected that the components of the EM field parallel to $\overline{\beta}$ are not changed by the Lorentz transform. Therefore, equation (5.7) must be rewritten. At this level, pure imaginary components and pure real components, either in the time domain or the space domain, are components of different vectors and so we need to separate out the two different types of vector in order to simplify things. Consequently, we separate out the two 4-vectors to get

$$\mathcal{L}_1(iL\hat{l}+\overline{E}+i\overline{B})=i\gamma(\overline{\beta}\square\overline{B})]\hat{l}-\gamma(\overline{B}-\overline{\beta}\times\overline{E})$$

$$-\gamma(L-\overline{\beta}\square\overline{E})\hat{l}+i\gamma(\overline{E}-L\overline{\beta}+\overline{\beta}\times\overline{B})$$ (6.9)

Then the components parallel and perpendicular to the velocity for the first sub vector are

$$i\gamma\left(\beta B_{\square}\right)\hat{i} - \gamma\left(B_{\square}\right)\overline{p} - \gamma\left(B_{\perp}\hat{q} - \overline{\beta}\times\overline{E}\right) \ (6.10)$$

and for the second sub vector are

$$-\gamma\left(L - \beta E_{\square}\right)\hat{i} + i\gamma\left(E_{\square} - \beta L\right)\overline{p} + i\gamma\left(E_{\perp}\hat{q} + \overline{\beta}\times\overline{B}\right). \ (6.11)$$

Setting $L = 0$, we get

$$\gamma\left(\beta E_{\square}\right)\hat{i} + i\gamma\left(E_{\square}\right)\overline{p} + i\gamma\left(E_{\perp}\hat{q} + \overline{\beta}\times\overline{B}\right) \ (6.12)$$

squaring and unsquaring, we get finally, for the first sub vector

$$-\gamma^2\beta^2 B_{\square}^2 + \gamma^2 B_{\square}^2 + [\gamma(B_{\perp}\hat{q} - \overline{\beta}\times\overline{E})]^2$$

$$B_{\square}\overline{p} + \gamma(B_{\perp}\hat{q} - \overline{\beta}\times\overline{E}) \qquad (6.13)$$

and for the second sub vector

$$\gamma^2\beta^2 E_{\square}^2 - \gamma^2 E_{\square}^2 + [i\gamma(E_{\perp}\hat{q} + \overline{\beta}\times\overline{B})]^2$$

$$iE_{\square}\overline{p} + i\gamma(E_{\perp}\hat{q} + \overline{\beta}\times\overline{B}) \qquad .(6.14)$$

Equations (6.13) and (614) are the correct results for Lorentz transformation of the E-M field.

In order to emphasize one of the deficiencies of the Conventional "Generalized" Lorentz transform, a diversion is made here to compare it to the Lorentz transformation developed by the author on the basis of the $\square$ matrix. The Conventional "Generalized" Lorentz transform will be designated by $\mathcal{LC}$ and the author's Lorentz transform, designated by $\mathcal{L}$, in the following. The conventional generalized Lorentz transform is taken from Reference 6.-.Morse and Feshbach, "Methods of Theoretical Physics", Vol. I, p. 95, McGraw-Hill, N.Y. (1953).

$\mathcal{LC} =$

$$
\begin{array}{cc}
\cosh\alpha & \sinh\alpha\sin\theta\cos\phi \\
\sinh\alpha\sin\theta\cos\phi & 1+(\cosh\alpha-1)\sin^2\theta\cos^2\phi \\
\sinh\alpha\sin\theta\sin\phi & (\cosh\alpha-1)\sin^2\theta\cos\phi\sin\phi \\
\sinh\alpha\cos\theta & (\cosh\alpha-1)\sin\theta\cos\theta\cos\phi
\end{array}
\qquad (6.15)
$$

$$
\begin{array}{cc}
\sinh\alpha\sin\theta\sin\phi & \sinh\alpha\cos\theta \\
(\cosh\alpha-1)\sin^2\theta\cos\phi\sin\phi & (\cosh\alpha-1)\sin\theta\cos\theta\cos\phi \\
1+(\cosh\alpha-1)\sin^2\theta\sin^2\phi & (\cosh\alpha-1)\sin\theta\cos\theta\sin\phi \\
(\cosh\alpha-1)\sin\theta\cos\theta\sin\phi & 1+(\cosh\alpha-1)\cos^2\theta
\end{array}
$$

where $\beta = \tanh\alpha$, $\sinh\alpha = \gamma\beta$, $\cosh\alpha = \gamma$, and, hence, $\beta_x = \beta\sin\theta\cos\phi$, $\beta_y = \beta\sin\theta\sin\phi$, $\beta_z = \beta\cos\theta$, and $\beta^2 = \beta_x^2 + \beta_y^2 + \beta_z^2$. Substituting these quantities into the matrix, one gets

$\mathcal{LC} =$

$$
\begin{array}{cc}
\gamma & \gamma\beta\sin\theta\cos\phi \\
\gamma\beta\sin\theta\cos\phi & 1+(\gamma-1)\sin^2\theta\cos^2\phi \\
\gamma\beta\sin\theta\sin\phi & (\gamma-1)\sin^2\theta\cos\phi\sin\phi \\
\gamma\beta\cos\theta & (\gamma-1)\sin\theta\cos\theta\cos\phi
\end{array}
\qquad (6.16)
$$

$$
\begin{array}{cc}
\gamma\beta\sin\theta\sin\phi & \gamma\beta\cos\theta \\
(\gamma-1)\sin^2\theta\cos\phi\sin\phi & (\gamma-1)\sin\theta\cos\theta\cos\phi \\
1+(\gamma-1)\sin^2\theta\sin^2\phi & (\gamma-1)\sin\theta\cos\theta\sin\phi \\
(\gamma-1)\sin\theta\cos\theta\sin\phi & 1+(\gamma-1)\cos^2\theta
\end{array}
$$

$$
\mathcal{LC} =
\begin{bmatrix}
\gamma & \gamma\beta_x & \gamma\beta_y & \gamma\beta_z \\
\gamma\beta_x & 1 & 0 & 0 \\
\gamma\beta_y & 0 & 1 & 0 \\
\gamma\beta_z & 0 & 0 & 1
\end{bmatrix}
+ \frac{(\gamma-1)}{\beta^2}
\begin{bmatrix}
0 & 0 & 0 & 0 \\
0 & \beta_x^2 & \beta_x\beta_y & \beta_x\beta_z \\
0 & \beta_y\beta_x & \beta_y^2 & \beta_y\beta_z \\
0 & \beta_z\beta_x & \beta_z\beta_y & \beta_z^2
\end{bmatrix}
\quad (6.17)
$$

Or

$$\mathcal{LC} = \begin{bmatrix} \gamma & \gamma\beta_x & \gamma\beta_y & \gamma\beta_z \\ \gamma\beta_x & 0 & 0 & 0 \\ \gamma\beta_y & 0 & 0 & 0 \\ \gamma\beta_z & 0 & 0 & 0 \end{bmatrix} + \frac{\gamma}{\beta^2}\begin{bmatrix} 0 & 0 & 0 & 0 \\ 0 & \beta_x^2 & \beta_x\beta_y & \beta_x\beta_z \\ 0 & \beta_y\beta_x & \beta_y^2 & \beta_y\beta_z \\ 0 & \beta_z\beta_x & \beta_z\beta_y & \beta_z^2 \end{bmatrix} +$$

$$\begin{bmatrix} 0 & 0 & 0 & 0 \\ 0 & 1 & 0 & 0 \\ 0 & 0 & 1 & 0 \\ 0 & 0 & 0 & 1 \end{bmatrix} - \begin{bmatrix} 0 & 0 & 0 & 0 \\ 0 & \alpha_x^2 & \alpha_x\alpha_y & \alpha_x\alpha_z \\ 0 & \alpha_y\alpha_x & \alpha_y^2 & \alpha_y\alpha_z \\ 0 & \alpha_z\alpha_x & \alpha_z\alpha_y & \alpha_z^2 \end{bmatrix} \qquad (6.18)$$

where $\alpha_x = \sin\theta\cos\phi$, $\alpha_y = \sin\theta\sin\phi$, and $\alpha_z = \cos\theta$. The reader will recall from Chapter II, page 2, equations (2.2) through (2.7), that the three dimensional matrices which are the second and fourth terms of the above equation (6.18) are periodic matrices of the A type and that these type matrices are squares of the X type matrices. That is

$$\begin{bmatrix} 0 & -\beta_z & \beta_y \\ \beta_z & 0 & -\beta_x \\ -\beta_y & \beta_x & 0 \end{bmatrix}^2 = \begin{bmatrix} \beta_x^2 & \beta_x\beta_y & \beta_x\beta_z \\ \beta_y\beta_x & \beta_y^2 & \beta_y\beta_z \\ \beta_z\beta_x & \beta_z\beta_y & \beta_z^2 \end{bmatrix} - \beta^2 I \quad (6.19) \; and$$

$$\begin{bmatrix} 0 & -\alpha_z & \alpha_y \\ \alpha_z & 0 & -\alpha_x \\ -\alpha_y & \alpha_x & 0 \end{bmatrix}^2 = \begin{bmatrix} \alpha_x^2 & \alpha_x\alpha_y & \alpha_x\alpha_z \\ \alpha_y\alpha_x & \alpha_y^2 & \alpha_y\alpha_z \\ \alpha_z\alpha_x & \alpha_z\alpha_y & \alpha_z^2 \end{bmatrix} - \alpha^2 I \quad (6.20)$$

Note that $\alpha^2 = 1$.

The author's Lorentz transform matrix is

$$\mathcal{L} = \begin{bmatrix} \gamma & -i\gamma\beta_x & -i\gamma\beta_y & -i\gamma\beta_z \\ i\gamma\beta_x & \gamma & i\gamma\beta_z & -i\gamma\beta_y \\ i\gamma\beta_x & -i\gamma\beta_z & \gamma & i\gamma\beta_x \\ i\gamma\beta_z & i\gamma\beta_y & -i\gamma\beta_x & \gamma \end{bmatrix} = \gamma\begin{bmatrix} 1 & -i\gamma\bar{\beta} \\ i\gamma\bar{\beta} & 1 \end{bmatrix} \quad (6.21)$$

In comparison with $\mathcal{LC}$, the conventional generalized Lorentz transform matrix, $\mathcal{L}$ is quite simple, even streamlined, easy to use, and has in the spatial part of the matrix a square root quality. It has the outstanding property of yielding the correct Lorentz transform of

44

all 4-vectors, especially for the electromagnetic field, which $\mathcal{LC}$ appears not to do and which ought to be the one thing that a Lorentz transform should do.

It is difficult to see how the cumbersome $\mathcal{LC}$ matrix can be applied to the E-M field vectors, because it transforms the electric field in exactly the same way as the magnetic field and does not mix the two fields as required. Although there is a relationship between $\mathcal{LC}$ and $\mathcal{L}$, this relationship is rather cumbersome itself.

To round out this section on electromagnetism, the E-M field due to a charge in arbitrary motion is calculated by a suitable D_1 matrix. Before doing so, we describe a compact way of writing N matrices and matrix equations in pseudo two dimensional form. This procedure greatly simplifies the computations and reduces the labor involved. The basic idea is to group the elements of the N matrix into a single time element and a 3-D spatial vector. For example, the N matrix is written as

$$\begin{pmatrix} d & \overline{a} \\ -\overline{a} & d \end{pmatrix},$$ where $\overline{a} = a\hat{i} + b\hat{j} + c\hat{k}$. Operation of this matrix on a

4-vector, $^4\overline{u} = s\hat{l} + x\hat{i} + y\hat{j} + z\hat{k} = s\hat{l} + \overline{x}$, is written as

$$\begin{pmatrix} d & \overline{a} \\ -\overline{a} & d \end{pmatrix}\begin{pmatrix} s\hat{l} \\ \overline{x} \end{pmatrix} = \begin{pmatrix} \left(ds + \overline{a}\square\overline{x}\right)\hat{l} \\ -\overline{a}s + d\overline{x} + \overline{a}\times\overline{x} \end{pmatrix}$$ where $\overline{x} = x\hat{i} + y\hat{j} + z\hat{k}$. Note

that the dot and cross products are always used exactly as shown here. As a further example, a matrix equation involving the D_1^* matrix and the electromagnetic potential is given below, which is equivalent to equation (5.1) above, thus

$$\begin{pmatrix} -\dfrac{i}{c}\dfrac{\partial}{\partial t} & \nabla \\ -\nabla & -\dfrac{i}{c}\dfrac{\partial}{\partial t} \end{pmatrix}\begin{pmatrix} i\phi\hat{l} \\ \overline{A} \end{pmatrix} = \begin{pmatrix} \left(\dfrac{1}{c}\dfrac{\partial\phi}{\partial t} + \nabla\square\overline{A}\right)\hat{l} \\ -i\nabla\phi - \dfrac{i}{c}\dfrac{\partial\overline{A}}{\partial t} + \nabla\times\overline{A} \end{pmatrix}.$$

I now return to the calculation of the E-M field of a single charge moving with arbitrary velocity. In this case, retardation must be taken into account. To get the potentials, the "infinitesimal"or charge Lorentz transform is simply modified by multiplying by the retardation factor, $\dfrac{1}{\kappa}$, where $\kappa = \hat{r}\square(\hat{r} - \bar{\beta}) = 1 - \beta_r$, thus

$$\frac{1}{\kappa}\begin{pmatrix} 1 & i\bar{\beta} \\ -i\bar{\beta} & 1 \end{pmatrix}\begin{pmatrix} i\dfrac{e}{r}\hat{i} \\ 0 \end{pmatrix} = \begin{pmatrix} i\dfrac{e}{\kappa r}\hat{i} \\ \dfrac{e\bar{\beta}}{\kappa r} \end{pmatrix}$$

Which yields for an elementary particle the well-known Lienard-Wiechert potentials.

To form the correct D_1^* matrix, we must use differential elements which take into account the finite velocity of propagation of the fields, i.e. the retardation effects. The correct elements are (see for example Reference 7 - W. K. H. Panofsky and M. Phillips, *Classical Electricity and Magnetism,* Addison-Wesley, Cambridge (1955), pp. 297 et seq.):

$$-\frac{i}{c}\frac{\partial}{\partial t'} \rightarrow -\frac{i}{\kappa c}\frac{\partial}{\partial t}$$

$$\nabla' \rightarrow \nabla - \frac{\bar{r}}{\kappa c r}\frac{\partial}{\partial t}$$, which give by simple substitution the right

$\square$ matrix. Operation with the D_1^* matrix on the Lienard-Wiechert potentials yields

$$\begin{pmatrix} -\dfrac{i}{\kappa c}\dfrac{\partial}{\partial t} & \nabla - \dfrac{\hat{r}}{\kappa c}\dfrac{\partial}{\partial t} \\ -\nabla + \dfrac{\hat{r}}{\kappa c}\dfrac{\partial}{\partial t} & -\dfrac{i}{\kappa c}\dfrac{\partial}{\partial t} \end{pmatrix}\begin{pmatrix} i\dfrac{e}{\kappa r}\hat{i} \\ \dfrac{e\bar{\beta}}{\kappa r} \end{pmatrix} =$$

$$e\begin{pmatrix} \left(\dfrac{1}{\kappa c}\dfrac{\partial}{\partial t}\left(\dfrac{1}{\kappa r} \right) + \nabla\square\left(\dfrac{\bar{\beta}}{\kappa r} \right) - \dfrac{\hat{r}}{\kappa c}\dfrac{\partial}{\partial t}\left(\dfrac{\bar{\beta}}{\kappa r} \right) \right)\hat{i} \\ i\left(-\nabla\left(\dfrac{1}{\kappa r} \right) + \dfrac{\hat{r}}{\kappa c}\dfrac{\partial}{\partial t}\left(\dfrac{1}{\kappa r} \right) - \dfrac{1}{\kappa c}\dfrac{\partial}{\partial t}\left(\dfrac{\bar{\beta}}{\kappa r} \right) \right) + \nabla\times\left(\dfrac{\bar{\beta}}{\kappa r} \right) - \dfrac{\hat{r}}{\kappa c}\times\dfrac{\partial}{\partial t}\left(\dfrac{\bar{\beta}}{\kappa r} \right) \end{pmatrix}$$

$$= e\left(\begin{array}{c}\left[\dfrac{-\beta^2+\beta_r-\kappa\beta_r+\kappa\beta^2+\beta_r\left(\beta^2-\beta_r\right)}{\kappa^3 r^2}+\dfrac{\hat{r}\square\dot{\overline{\beta}}\left(1-\kappa-\beta_r\right)}{\kappa^3 cr}\right]\hat{i}\\[3em] i\left[\dfrac{\kappa\left(\hat{r}-\overline{\beta}\right)-\hat{r}\left(\beta^2-\beta_r\right)+\overline{\beta}\left(\beta^2-\beta_r\right)}{\kappa^3 r^2}+\dfrac{\hat{r}\left(\hat{r}\square\dot{\overline{\beta}}\right)-\kappa\dot{\overline{\beta}}-\overline{\beta}\left(\hat{r}\square\dot{\overline{\beta}}\right)}{\kappa^3 cr}\right]\\[3em] +\left[\dfrac{\kappa\left(\hat{r}-\overline{\beta}\right)\times\overline{\beta}-\left(\hat{r}\times\overline{\beta}\right)\left(\beta^2-\beta_r\right)}{\kappa^3 r^2}+\dfrac{-\kappa\left(\hat{r}\times\dot{\overline{\beta}}\right)-\left(\hat{r}\square\dot{\overline{\beta}}\right)\left(\hat{r}\times\overline{\beta}\right)}{\kappa^3 cr}\right]\end{array}\right)$$

$$= e\left(\begin{array}{c}[0]\hat{i}\\[3em] i\left[\dfrac{\left(1-\beta^2\right)\left(\hat{r}-\overline{\beta}\right)}{\kappa^3 r^2}+\dfrac{\left(\hat{r}\square\dot{\overline{\beta}}\right)\left(\hat{r}-\overline{\beta}\right)-\kappa\dot{\overline{\beta}}}{\kappa^3 cr}\right]\\[3em] +\left[\dfrac{\kappa\left(\hat{r}\times\overline{\beta}\right)\left(1-\beta^2\right)}{\kappa^3 r^2}-\dfrac{\left(\hat{r}\times\overline{\beta}\right)\left(\hat{r}\square\dot{\overline{\beta}}\right)+\kappa\left(\hat{r}\times\dot{\overline{\beta}}\right)}{\kappa^3 cr}\right]\end{array}\right)=\left(\begin{array}{c}0\hat{i}\\ i\overline{E}+\overline{B}\end{array}\right),$$

where $\kappa=1-\beta_r$ and all quantities in brackets are evaluated at the retarded time. When multiplied by $-i$, this is the correct result for arbitrary motion of a point charge. Notice, however, that the Lorentz term, or fourth component, is identically and automatically equal to zero. There is now a strong case for the absence of a fourth component to the E-M field; because not even an arbitrarily moving charge can produce such a component.

47

Chapter VII. Applications of ☐ Operators in Quantum Mechanics

Equation Section (Next)

Quantum Mechanics has the advantage, over most of the other physics theories, of making it possible to calculate physical quantities to extreme precision, while, at the same time, explaining nothing about them. Quantum mechanics is amenable to a peculiar type of development by means of, algorithms, such as quantization of equations by commutator brackets. The theoretical structure of QM is fairly transparent, but it is often extremely difficult, if not impossible, to interpret. At this time in the development of physics, we seem to be faced, at worst, with generation of a set of algorithms without much hope of really understanding how nature works and, at best, with a monumental task of trying to develop a basis whereby human understanding can play a more direct part in this development.

Quantum Mechanics involves some of the more arcane and non-intuitive concepts of physics, such as indefiniteness, chance and probability, discontinuous change, entanglement, and exclusion to name a few. Indeed, the weird concept of wave-particle duality seems to be the basic feature of quantum mechanics – more basic than the uncertainty principle. One must consider an "elementary particle" to have the attributes of *both* a particle and a wave. An electron, for example, is localized or not localized in an odd sort of way apparently determined by the uncertainty principle. It has a wave nature that is not directly observed. A diffraction or interference pattern is discerned only after a relatively large number of electrons have passed through suitable slits. For free photons, on the other hand, the wave aspect is directly apparent, but the particle aspect is not, unless microscopic examination of interference or diffraction patterns is made. If an experiment is set up to measure the wave aspect of either photon or electron, then the results are consistent with a wave; and, if the experiment is set up to measure the particle aspect of either, the results are consistent with a particle. Part, if not all, of this is tied up in the process of detection (i.e. how we measure whether something is there or not). For example, if we try to detect light by the photoelectric effect, it appears to be made up of photons; and, if we try to detect light by a parametric oscillator, we will detect the frequency or wavelength of the light. Classical intuition is not very useful; our classical intuition appears in many cases to be wrong.

What are the measurable properties that distinguish waves and particles? In its simplest form, a free particle is usually imagined classically to be a localized bit of energy or mass for which a center of mass can be relatively accurately measured and which propagates with a single constant group velocity at the center of mass. In its simplest form, a wave is usually thought of classically as a non localized disturbance in a continuous medium that propagates with a least a phase velocity and, if even slightly localized, a group velocity also. In the first place, is it possible that there may not actually be any truly distinguishing properties, even from a classical standpoint? We can see that even classically the distinction is not so clear cut. The fundamental distinction seems to be in the number of particles that can be considered to be involved in a particular situation, i.e. of the number of interacting entities with internal rigidity involved (all entities possessing extent, of course, must undergo relativistic deformation). At any given instant of time then, a classical particle is "rigid", has a definite position, and has only one momentum or velocity vector which is located at its center of mass, while a classical wave is not rigid and has different velocity or momentum vectors at a great number of points in its volume. As a consequence, a wave tends to spread out with time while a particle remains localized. A wave seems to require the fiction of a continuous medium while a particle seems not to.

(Note added in proof: It has very recently come to my attention that a very clever experiment by Shahriar Afshar (see New Scientist, 17 February 2007, p. 13, "Quantum rebel wins over doubters"), in which a wire screen was placed to cover the minima in the diffraction pattern of two slits indicates that both the diffraction of and particle nature of photons are detected simultaneously. Consequently, Bohr's Complimentarity Principle has been significantly challenged.)

Can one construct a theory of an entity that has the attributes of both a particle and a wave? A possible approach begins with fundamentals and generalizes the mathematics. The relativistic energy momentum equation is the most fundamental equation of the motion of an elementary particle that we have at hand. In the previous Chapters, I have endeavored to show that the $\square$ operators are important in that they can be used to develop a lot of physics. In particular the Lorentz transformation operator, $\mathcal{L}_l$, and the differential operator, D ,should also be fundamental. Consider again equation (5.26) for operation of the D_1^* matrix on the relativistic energy-momentum vector. This procedure yields a more general force equation,

$$D_1^*\left(i\mathcal{E}\hat{1}+c\overline{p}\right)=\left(\frac{1}{c}\frac{\partial\mathcal{E}}{\partial t}+c\nabla\square\overline{p}\right)\hat{1}-i\left(\nabla\mathcal{E}+\frac{\partial\overline{p}}{\partial t}\right)+c\nabla\times\overline{p}.$$

The first component on the right is a kind of continuity statement for energy and momentum. The second component is the force-momentum relationship and includes force fields, if the total energy includes the potential energies of the force fields. The last component is the vorticity, if any, of the momentum. These components apply to a classical particle and to a classical wave, provided suitable Dirac delta functions and suitable summations are used to deal with singularities and suitable energy density functions and suitable momentum densities are used, respectively.

The first component looks like a momentum-energy current. If there are no sources or sinks, we set this component equal to zero, thus

$$\frac{1}{c}\frac{\partial\mathcal{E}}{\partial t}+c\nabla\square\overline{p}=0.$$

It only remains to apply Gauss' theorem to show that energy density or energy and momentum density or momentum are conserved. Equation (5.26) must include the rest mass energy of the particle. Therefore with this additional modification, the equation will work for particles as well as waves.

For the second component, provided the total energy includes the potential energies due to all forces acting on the particle (in so far as that can be done in a physically meaningful way) and the partial time derivative is replaced by the total derivative with respect to proper time, the correct equation for the force on a particle in the limit

$\beta \square \; 1$ (see Ref. 3) is obtained, but only if we also set this component equal to zero, thus

$$\frac{d\overline{p}}{dt} + \nabla V = 0 \quad \text{where} \quad \nabla\left(\gamma m_0 c^2\right) = 0, \text{ i.e. the particle exerts no}$$

force on itself, e.g. by changing its rest mass. (It should be apparent that all forces, such as dissipative forces, which cannot be represented by the gradient of a potential, have been set to zero.) This also applies for a wave, if we use energy densities and energy flow densities.

The third component, if made into an equation, would be a condition on the vorticity of the momentum of the wave or particle. In the case of a structureless particle the vorticity of the momentum would necessarily be zero; and in the case of a wave it could certainly be finite.

So can we find a mathematical object for which the energy and momentum is such that our three equations are satisfied and which has the aspects of both a particle and a wave? The best clue comes from the third equation (vorticity of the momentum), i.e. the point particle aspect would require $\nabla \times \overline{p} = 0$, and the wave aspect might require $\nabla \times \overline{p} \neq 0$. To satisfy both of these conditions at once we could make $\overline{p} = -i\hbar \nabla \psi + \nabla \times \overline{u}$, where $\psi = \psi(\overline{r}, t)$ and $\overline{u} = \overline{u}(\overline{r}, t)$; so that part of the momentum satisfies $\nabla \times \overline{p} = 0$ and the other part satisfies $\nabla \times \overline{p} \neq 0$, specifically, $\nabla \times \overline{p} = \nabla \times \left(\nabla \times \overline{u}\right)$. Where we have taken advantage of Helmholtz's theorem to maintain as much generality as possible and of our prior knowledge that quantum mechanics requires that $-i\hbar$ multiplies the gradient term (a small cheat). Indiscriminately we plug our momentum term into all of the three equations to get,

$$\frac{1}{c}\frac{\partial \mathcal{E}}{\partial t} - i\hbar c \nabla^2 \psi = C_0 m_0 c^2 \psi$$

$$\nabla \mathcal{E} - i\hbar \left[\frac{\partial(\nabla \psi)}{\partial t} + \frac{\partial(\nabla \times \overline{u})}{\partial t} \right] = 0$$

$$\nabla \times \overline{p} = \nabla \times \left(\nabla \times \overline{u}\right) = \nabla\left(\nabla \square \overline{u}\right) - \nabla^2 \overline{u}.$$

51

From the second equation we see that if $\frac{\partial \overline{u}}{\partial t} = 0$, then

$$\nabla\left(\mathcal{E} - i\hbar\frac{\partial \psi}{\partial t}\right) = 0 \quad or$$

$\mathcal{E} = i\hbar\frac{\partial \psi}{\partial t} + C_1$ where C_1 is a constant w.r.t. space coordinates, at least. Substituting this result into the first equation yields

$\frac{i\hbar}{c}\frac{\partial^2 \psi}{\partial t^2} - i\hbar c\nabla^2\psi = C_0 m_0 c^2\psi - \frac{1}{c}\frac{\partial C_1}{\partial t}$ we now have a new set of equations: namely,

$$\frac{i\hbar}{c}\frac{\partial^2 \psi}{\partial t^2} - i\hbar c\nabla^2\psi = C_0 m_0 c^2\psi - \frac{1}{c}\frac{\partial C_1}{\partial t}$$

$$\mathcal{E} = i\hbar\frac{\partial \psi}{\partial t} + C_1$$

$$\overline{p} = -i\hbar\nabla\psi + \nabla\times\overline{u}$$

$$\nabla C_1 = 0$$

$$\frac{\partial \overline{u}}{\partial t} = 0$$

$$\nabla\times\overline{p} = \nabla\times\left(\nabla\times\overline{u}\right) = \nabla\left(\nabla\square\overline{u}\right) - \nabla^2\overline{u}.$$

We are still free, I believe, to select C_0, C_1, and $\nabla\square\overline{u}$ without significant loss of generality. Let $C_0 = -i\frac{m_0 c}{\hbar}$ and, because C_1 merely determines the zero of the total energy, let $C_1 = 0$ (this is not a problem if there are no accelerations or the velocities involved are small compared to c). Then, the first equation immediately above becomes the so called Klein-Gordon equation,

$$-\hbar^2\frac{\partial^2 \psi}{\partial t^2} + \hbar^2 c^2\nabla^2\psi = m_0^2 c^4\psi .$$

This is an important result of our new approach to understanding the wave-particle duality of quantum mechanics. This a wave equation for something that has a rest mass, m_0. Also, we have obtained the quantum mechanical energy and momentum operators in terms of differentials without an ad hoc assumption.

Continuing in the same spirit, we approach solving the Klein-Gordon equation, that is, we try to find a function $\psi\left(\bar{r}, ct\right)$ that is both wavelike (oscillatory and spread out) and particle-like (non-oscillatory and localized) at the same time as far as possible. The argument of this function should also be Lorentz covariant. We find almost immediately that our Lorentz transformed 4-D position vector is a capital starting point, and the scalar product of that with a 4-D propagation vector ($^4\bar{k} = ik_4\hat{l} + i\bar{k}$) gives a suitable argument for the function ψ. The result is

$$\psi = \psi_0 \exp\left[-ik_4\gamma\left(ct - \bar{\beta}\Box\bar{r}\right) - \gamma\left(\bar{k}\Box\bar{r} - \bar{k}\Box\bar{\beta}ct\right) + i\gamma\left(\bar{k}\times\bar{\beta}\right)\Box\bar{r}\right] \quad .(7.1)$$

The first term of the argument is oscillatory and travels in the $\bar{\beta}$ direction with a phase velocity of magnitude c/β. The second term confines the wavefunction to a localized region of space of the order of $1/\gamma k$ in size in all space directions, the center of which travels in the $\bar{\beta}$ direction with the group velocity $c\bar{\beta}$. For the time being, we will ignore the part of the momentum given above by $\nabla\times\bar{u}$, i.e. we take $\nabla\times\bar{u} = 0$, so that $\bar{p} = -i\hbar\nabla\psi$ and $\mathcal{E} = i\hbar\dfrac{\partial\psi}{\partial t}$ and operate only with these elementary operators in order to determine if our trial ψ is a solution of the Klein-Gordon equation, thus

$$\nabla\psi = \left(ik_4\gamma\bar{\beta} - \gamma\bar{k} + i\gamma\bar{k}\times\bar{\beta}\right)\psi \quad and$$

$$\frac{\partial\psi}{\partial t} = -\left(i\gamma k_4 c - \gamma\bar{k}\Box\bar{\beta}c\right)\psi .$$

In general,

$$\nabla\psi = \left[-\gamma\left(k_x - ik_y\beta_z + ik_z\beta_y - ik_4\beta_x\right)\hat{i} - \gamma\left(k_y - ik_z\beta_x + ik_x\beta_z - ik_4\beta_y\right)\hat{j} - \gamma\left(k_z - ik_x\beta_y + ik_y\beta_x - ik_4\beta_z\right)\hat{k}\right]\psi$$

$$(7.2) \; and$$

$$\frac{\partial\psi}{\partial t} = -\gamma c\left(ik_4 - k_x\beta_x - k_y\beta_y - k_z\beta_z\right)\psi .(7.3)$$

Therefore,

$$-\hbar^2 \frac{\partial^2 \psi}{\partial t^2} + \hbar^2 c^2 \nabla^2 \psi = \gamma^2 \hbar^2 c^2 \left[k_x^2 \left(1 - \beta^2\right) + k_y^2 \left(1 - \beta^2\right) + k_z^2 \left(1 - \beta^2\right) + k_4^2 \left(1 - \beta^2\right) \right]$$

$$= \hbar^2 c^2 \left(k_x^2 + k_y^2 + k_z^2 + k_4^2 \right) \psi = \hbar^2 c^2 \left(\overline{k}^2 + k_4^2 \right) \psi = m_0^2 c^4 \psi$$

Consequently, ψ is a solution provided that $\overline{k}^2 + k_4^2 = \dfrac{m_0^2 c^2}{\hbar^2}$.

This is a remarkable result, for we get, making the simplest assumptions that the components of the propagation vector are all equal in magnitude (= k) and that the wavefunction is somewhat localized, the result,

$$k^2 = \frac{m_0^2 c^2}{4\hbar^2} \quad .$$

The same as if the wave-particle had spin of one half. However, this is clearly a result of using 4-vectors in deriving the argument of ψ (4-D propagation and 4-D position vectors).

Needless to say, if $\overline{k}^2 + k_4^2 = 0$, i.e. the rest mass is zero, then ψ is a solution of the Wave Equation. We take, as the normal form of the argument of ψ, shown again below for reference, the quantity

$$\left[-ik_4 \gamma \left(ct - \overline{\beta} \Box \overline{r} \right) - \gamma \left(\overline{k} \Box \overline{r} - \overline{k} \Box \overline{\beta} ct \right) + i\gamma \left(\overline{k} \times \overline{\beta} \right) \Box \overline{r} \right].$$

The time inverted form of the argument is

$$\left[ik_4 \gamma \left(ct - \overline{\beta} \Box \overline{r} \right) - \gamma \left(\overline{k} \Box \overline{r} - \overline{k} \Box \overline{\beta} ct \right) - i\gamma \left(\overline{k} \times \overline{\beta} \right) \Box \overline{r} \right] \quad \text{and} \quad \text{the space}$$

inverted form of the argument is

$$\left[-ik_4 \gamma \left(ct - \overline{\beta} \Box \overline{r} \right) + \gamma \left(\overline{k} \Box \overline{r} - \overline{k} \Box \overline{\beta} ct \right) + i\gamma \left(\overline{k} \times \overline{\beta} \right) \Box \overline{r} \right].$$

All of these arguments yield solutions of the Klein-Gordon equation or the Wave Equation, provided that $\overline{k}^2 + k_4^2$ is non-zero or zero respectively. However, the energy and momentum are not what one might at first expect, thus, in the case of the normal, non-inverted, argument,

$$\overline{p} = \gamma \hbar k_4 \overline{\beta} + i\gamma \hbar \overline{k} - i\gamma \hbar \overline{k} \times \overline{\beta} = \frac{\gamma m_0 c \overline{\beta}}{2} + \frac{i\gamma m_0 c \overline{k}}{2k} - \frac{i\gamma m_0 c \overline{k} \times \overline{\beta}}{2k}$$

$$\varepsilon = \gamma \hbar k_4 c + i\gamma \hbar k_z \beta c = \frac{\gamma m_0 c^2}{2} + \frac{i\gamma m_0 c^2 \beta}{2} .$$

In addition to the normal energy and momentum being half of what they should be; there are the extraneous momentum and energy

$$\overline{p}_e = \frac{i\gamma m_0 c \overline{k}}{2k} - \frac{i\gamma m_0 c \overline{k} \times \overline{\beta}}{2k}$$

$$\mathcal{E}_e = \frac{i\gamma m_0 c^2 \overline{\beta}}{2}.$$

These quantities must, however, satisfy the relativistic energy-momentum relationship, $\mathcal{E}^2 = c^2 p^2 + m_0^2 c^4$, which, in fact, altogether they do. For clarity and simplicity of discussion, a plane wave is temporarily assumed, so that $k_x = k_y = 0$, and

$$\overline{p}_e = \frac{i\gamma m_0 c \overline{k}}{2}$$

$\mathcal{E}_e = \dfrac{i\gamma m_0 c^2 \overline{\beta}}{2}$ are the extraneous momentum and energy and

$$\overline{p}_n = \frac{\gamma m_0 c \overline{\beta}}{2}$$

$\mathcal{E}_n = \dfrac{\gamma m_0 c^2}{2}$ are the normal energy and momentum. Thus, one can see that the extraneous energy, $\mathcal{E}_e$ is, however, just the amount of energy needed to produce $\overline{p}_n$ and is, therefore, the energy of the momentum or, in other words, the kinetic energy plus the rest energy. The extraneous momentum, $\overline{p}_e$, is the 4th component of the momentum and $\overline{p}_n$ is the space component of the momentum. $\mathcal{E}_n$ is, of course, the total energy.

It is enlightening to rewrite the energy and momentum in terms of the momentum 4-vector, thus

$$^4\overline{p} = \gamma\hbar\left(ik_x + k_y\beta_z - k_z\beta_y + k_4\beta_x\right)\hat{i} + \gamma\hbar\left(ik_y + k_z\beta_x - k_x\beta_z + k_4\beta_y\right)\hat{j}$$
$$+\gamma\hbar\left(ik_z + k_x\beta_y - k_y\beta_x + k_4\beta_z\right)\hat{k} + \gamma\hbar\left(ik_4 - k_x\beta_x - k_y\beta_y - k_z\beta_z\right)\hat{l}$$

(7.4) or

$$^4\overline{p} = i\gamma\hbar \begin{bmatrix} 1 & 0 & 0 & 0 \\ 0 & 1 & 0 & 0 \\ 0 & 0 & 1 & 0 \\ 0 & 0 & 0 & 1 \end{bmatrix} {}^4\overline{k} - \gamma\hbar\beta_x \begin{bmatrix} 0 & 1 & 0 & 0 \\ -1 & 0 & 0 & 0 \\ 0 & 0 & 0 & -1 \\ 0 & 0 & 1 & 0 \end{bmatrix} {}^4\overline{k}$$

$$-\gamma\hbar\beta_y \begin{bmatrix} 0 & 0 & 1 & 0 \\ 0 & 0 & 0 & 1 \\ -1 & 0 & 0 & 0 \\ 0 & -1 & 0 & 0 \end{bmatrix} {}^4\overline{k} - \gamma\hbar\beta_z \begin{bmatrix} 0 & 0 & 0 & 1 \\ 0 & 0 & -1 & 0 \\ 0 & 1 & 0 & 0 \\ -1 & 0 & 0 & 0 \end{bmatrix} {}^4\overline{k} \qquad (7.5)$$

where

$$^4\overline{k} = \begin{pmatrix} k_4 \\ k_x \\ k_y \\ k_z \end{pmatrix}$$

The 4-momentum can be written in terms of the $i\gamma_y = \Xi$, $\delta_0 = \Psi$, and $i\beta_y = Z$ spin matrices (they are also inversion matrices, see the equations on p. 14 above) derived from the N matrix by replacing all the a, b, and c elements in turn with $\sin\theta = \sin\frac{\pi}{2} = 1$. The representation here is not the same as that used by Bjorken and Drell, but the physics is essentially the same. One thus obtains the result,

$$^4\overline{p} = i\gamma\hbar I\,^4\overline{k} - \gamma\hbar\left(\beta_x X + \beta_y Y + \beta_z Z\right){}^4\overline{k}.$$

Hence, if we write the matrices Ξ, Ψ, and Z in the form of a 3-vector, thus

$$\overline{A} = \hat{i}X + \hat{j}Y + \hat{k}Z ; \textit{ the 4-momentum can be written as}$$

$$^4\overline{p} = i\gamma\hbar I\,^4\overline{k} - \gamma\hbar\overline{\beta}\square\overline{A}\,^4\overline{k} = -\gamma\hbar\left(T + \overline{\beta}\square\overline{A}\right){}^4\overline{k}, \quad (7.6)$$

where $T = iI$. This result is reminiscent of Dirac's equation, especially because the matrices I, Ξ, Ψ, and Z have the following multiplication table:

	I	X	Y	Z
I	I	X	Y	Z
X	X	-I	Z	-Y
Y	Y	-Z	-I	X
Z	Z	Y	-X	-I

where I is the identity matrix. It is only necessary to put in the rest mass term and change the 4-velocity to a 4-D differential operator.

Further, the momentum equation can be written entirely in terms of 4-vectors, thus

$$^4\overline{p} = -\gamma\hbar\left(^4\overline{\beta}\square\,^4\overline{A}\right)^4\overline{k}, \text{ where}$$

$$^4\overline{A} = I\hat{l} + X\hat{i} + Y\hat{j} + Z\hat{k},$$

$$^4\overline{\beta} = i\hat{l} + \overline{\beta} \text{ and}$$

$$^4\overline{k} = k_4\hat{l} + \overline{k}.$$

A final brief digression is made to point out that

$$\gamma\left(^4\overline{\beta}\square\,^4\overline{A}\right) = \mathcal{L}_l, \text{ is the Lorentz transformation operator, rather}$$

surprisingly; so that $^4\overline{p} = -\hbar\mathcal{L}_l\,^4\overline{k}$. Consequently, the 4-momentum is $\hbar$ times the Lorentz transform of the propagation 4-vector.

Reexamination of the 4-momentum vector shows that the velocity components and the position coordinate components are tangled up (it is apparent that they are snarled up by the operation of the spin matrices), so that, if these are unsnarled, we might get the correct momentum components. Unsnarling begins by regrouping the momentum components with common velocity components (unsnarling could also be done by operation of the spin matrices a suitable number of times), thus

$$p_1 = \gamma\hbar\left(k_4\hat{i} + k_z\hat{j} - k_y\hat{k} - k_x\hat{l}\right)\beta_x$$

$$p_2 = \gamma\hbar\left(k_z\hat{i} + k_4\hat{j} + k_x\hat{k} - k_y\hat{l}\right)\beta_y$$

$$p_3 = \gamma\hbar\left(k_y\hat{i} - k_x\hat{j} + k_4\hat{k} - k_z\hat{l}\right)\beta_z \qquad (7.7)$$

$$p_l = \gamma\hbar\left(k_x\hat{i} + k_y\hat{j} + k_z\hat{k} + k_4\hat{l}\right)$$

Squaring each component separately, one gets

$$p_1^2 = \gamma^2 \hbar^2 \left(k_4^2 + k_z^2 + k_y^2 + k_x^2 \right) \beta_x^2 = \gamma^2 m_0^2 c^2 \beta_x^2$$

$$p_2^2 = \gamma^2 \hbar^2 \left(k_z^2 + k_4^2 + k_x^2 + k_y^2 \right) \beta_y^2 = \gamma^2 m_0^2 c^2 \beta_y^2$$

$$p_3^2 = \gamma^2 \hbar^2 \left(k_y^2 + k_x^2 + k_4^2 + k_z^2 \right) \beta_z^2 = \gamma^2 m_0^2 c^2 \beta_z^2$$

$$p_t^2 = \gamma^2 \hbar^2 \left(k_x^2 + k_y^2 + k_z^2 + k_4^2 \right) = \gamma^2 m_0^2 c^2$$

(7.8)

which results in the correct values of the momentum and energy

provided $k^2 = k_x^2 + k_y^2 + k_z^2 + k_4^2$ and $k^2 = \dfrac{m_0^2 c^2}{\hbar^2}$.

At this point, the next step would be to demonstrate that it is possible to obtain a linear relativistic wave equation similar to the Dirac equation, which gives the correct energy levels and line spreads for the hydrogen-like atom, but which also avoids snarling up the elements in the spin, velocity, and other matrices of the wave equation. This could possibly be accomplished by use of additional 4-D ☐ matrices or by 8-D ☐ matrices. Suitable such matrices can be chosen from the following.

4-D Matrices

$$i\gamma_y = X = \begin{pmatrix} 0 & 1 & 0 & 0 \\ -1 & 0 & 0 & 0 \\ 0 & 0 & 0 & -1 \\ 0 & 0 & 1 & 0 \end{pmatrix} \qquad \delta_0 = Y = \begin{pmatrix} 0 & 0 & 1 & 0 \\ 0 & 0 & 0 & 1 \\ -1 & 0 & 0 & 0 \\ 0 & -1 & 0 & 0 \end{pmatrix}$$

$$i\beta_y = Z = \begin{pmatrix} 0 & 0 & 0 & 1 \\ 0 & 0 & -1 & 0 \\ 0 & 1 & 0 & 0 \\ -1 & 0 & 0 & 0 \end{pmatrix}$$

That is from the matrices

$\alpha_x, i\alpha_y, \alpha_z, \alpha_0, \beta_x, i\beta_y, \beta_z, \beta_0, \gamma_x, i\gamma_y, \gamma_z, \gamma_0, \delta_x, i\delta_y, \delta_z,$ and δ_0

defined on page 60 below.

58

8-D Matrices

To begin with, two 8x8 matrices are generated from certain 8-D N matrices as follows:

$$\begin{pmatrix} d & a & b & c & h & e & f & g \\ -a & d & -c & b & -e & h & g & -f \\ -b & c & d & -a & -f & -g & h & e \\ -c & -b & a & d & -g & f & -e & h \\ -h & e & f & g & -d & -a & -b & -c \\ -e & -h & g & -f & a & -d & c & -b \\ -f & -g & -h & e & b & -c & -d & a \\ -g & f & -e & -h & c & b & -a & -d \end{pmatrix}^2 = 2d \begin{pmatrix} d & a & b & c & 0 & 0 & 0 & 0 \\ -a & d & -c & b & 0 & 0 & 0 & 0 \\ -b & c & d & -a & 0 & 0 & 0 & 0 \\ -c & -b & a & d & 0 & 0 & 0 & 0 \\ 0 & 0 & 0 & 0 & d & a & b & c \\ 0 & 0 & 0 & 0 & -a & d & -c & b \\ 0 & 0 & 0 & 0 & -b & c & d & -a \\ 0 & 0 & 0 & 0 & -c & -b & a & d \end{pmatrix}$$

$$-\left(a^2 + b^2 + c^2 + d^2 + e^2 + f^2 + g^2 + h^2\right)I$$

(7.9) where the first matrix on the right side of equation (7.9) is one of the matrices we will use, after minor modification, to construct a linear relativistic wave equation. We will call this matrix $\overline{\overline{\sigma}}$. The other matrix is generated by multiplying two N matrices together as follows:

$$\begin{pmatrix} d & a & b & c & h & e & f & g \\ -a & d & -c & b & -e & h & g & f \\ -b & c & d & -a & -f & g & h & e \\ -c & -b & a & d & -g & f & e & h \\ -h & e & f & g & d & a & b & c \\ -e & -h & g & f & -a & d & -c & b \\ -f & -g & -h & e & -b & c & d & -a \\ -g & f & -e & -h & -c & -b & a & d \end{pmatrix} \begin{pmatrix} -d & a & b & c & h & e & f & g \\ -a & -d & -c & b & -e & h & -g & f \\ -b & c & -d & -a & -f & g & h & -e \\ -c & -b & a & -d & -g & -f & e & h \\ -h & e & f & g & -d & a & b & c \\ -e & -h & -g & f & -a & -d & -c & b \\ -f & g & -h & -e & -b & c & -d & -a \\ -g & -f & e & -h & -c & -b & a & -d \end{pmatrix}$$

$$= 2h \begin{pmatrix} 0 & 0 & 0 & 0 & 0 & a & b & c \\ 0 & 0 & 0 & 0 & -a & 0 & -c & b \\ 0 & 0 & 0 & 0 & -b & c & 0 & -a \\ 0 & 0 & 0 & 0 & -c & -b & a & 0 \\ 0 & -a & -b & -c & 0 & 0 & 0 & 0 \\ a & 0 & c & -b & 0 & 0 & 0 & 0 \\ b & -c & 0 & a & 0 & 0 & 0 & 0 \\ c & b & -a & 0 & 0 & 0 & 0 & 0 \end{pmatrix} -$$

$$2q \begin{pmatrix} 0 & 0 & 0 & 0 & 1 & 0 & 0 & 0 \\ 0 & 0 & 0 & 0 & 0 & 1 & 0 & 0 \\ 0 & 0 & 0 & 0 & 0 & 0 & 1 & 0 \\ 0 & 0 & 0 & 0 & 0 & 0 & 0 & 1 \\ 1 & 0 & 0 & 0 & 0 & 0 & 0 & 0 \\ 0 & 1 & 0 & 0 & 0 & 0 & 0 & 0 \\ 0 & 0 & 1 & 0 & 0 & 0 & 0 & 0 \\ 0 & 0 & 0 & 1 & 0 & 0 & 0 & 0 \end{pmatrix}$$

$$-\left(a^2 + b^2 + c^2 + d^2 + e^2 + f^2 + g^2 + h^2\right)I \qquad (7.10)$$

where $q = (ae + bf + cg)$. The first matrix on the right side of equation (7.10) is the other matrix that we will use after minor modification. It will be called $\overline{\overline{\alpha}}$. Besides these two matrices, we will need a matrix $\overline{\overline{\beta}}$ obtained from the first generating matrix by setting d = 1 and all other elements equal to zero and we will need the 8-D identity matrix, of course. These matrices are displayed as follows:

$$
\overline{\overline{\alpha}} = \begin{pmatrix}
0 & 0 & 0 & 0 & 0 & \hat{i} & \hat{j} & \hat{k} \\
0 & 0 & 0 & 0 & -\hat{i} & 0 & -\hat{k} & \hat{j} \\
0 & 0 & 0 & 0 & -\hat{j} & \hat{k} & 0 & -\hat{i} \\
0 & 0 & 0 & 0 & -\hat{k} & -\hat{j} & \hat{i} & 0 \\
0 & -\hat{i} & -\hat{j} & -\hat{k} & 0 & 0 & 0 & 0 \\
\hat{i} & 0 & \hat{k} & -\hat{j} & 0 & 0 & 0 & 0 \\
\hat{j} & -\hat{k} & 0 & \hat{i} & 0 & 0 & 0 & 0 \\
\hat{k} & \hat{j} & -\hat{i} & 0 & 0 & 0 & 0 & 0
\end{pmatrix}
\qquad (7.11)
$$

$$
\overline{\overline{\sigma}} = i \begin{pmatrix}
0 & \hat{i} & \hat{j} & \hat{k} & 0 & 0 & 0 & 0 \\
-\hat{i} & 0 & -\hat{k} & \hat{j} & 0 & 0 & 0 & 0 \\
-\hat{j} & \hat{k} & 0 & -\hat{i} & 0 & 0 & 0 & 0 \\
-\hat{k} & -\hat{j} & \hat{i} & 0 & 0 & 0 & 0 & 0 \\
0 & 0 & 0 & 0 & 0 & \hat{i} & \hat{j} & \hat{k} \\
0 & 0 & 0 & 0 & -\hat{i} & 0 & -\hat{k} & \hat{j} \\
0 & 0 & 0 & 0 & -\hat{j} & \hat{k} & 0 & -\hat{i} \\
0 & 0 & 0 & 0 & -\hat{k} & -\hat{j} & \hat{i} & 0
\end{pmatrix}
\qquad (7.12)
$$

$$
\overline{\overline{\beta}} = \begin{pmatrix}
1 & 0 & 0 & 0 & 0 & 0 & 0 & 0 \\
0 & 1 & 0 & 0 & 0 & 0 & 0 & 0 \\
0 & 0 & 1 & 0 & 0 & 0 & 0 & 0 \\
0 & 0 & 0 & 1 & 0 & 0 & 0 & 0 \\
0 & 0 & 0 & 0 & -1 & 0 & 0 & 0 \\
0 & 0 & 0 & 0 & 0 & -1 & 0 & 0 \\
0 & 0 & 0 & 0 & 0 & 0 & -1 & 0 \\
0 & 0 & 0 & 0 & 0 & 0 & 0 & -1
\end{pmatrix}
\qquad (7.13)
$$

where a, b, and c are replaced by the unit vectors $\hat{i}$, $\hat{j}$, and $\hat{k}$ respectively. The vector components of $\overline{\overline{\alpha}}$ and $\overline{\overline{\beta}}$ are found by dot products with the unit vectors, thus

$$\overline{\overline{\alpha}}\Big|_x = \hat{i}\square\overline{\overline{\alpha}} = \overline{\overline{\alpha}}\square\hat{i} = \begin{pmatrix} 0 & 0 & 0 & 0 & 0 & 1 & 0 & 0 \\ 0 & 0 & 0 & 0 & -1 & 0 & 0 & 0 \\ 0 & 0 & 0 & 0 & 0 & 0 & 0 & -1 \\ 0 & 0 & 0 & 0 & 0 & 0 & 1 & 0 \\ 0 & -1 & 0 & 0 & 0 & 0 & 0 & 0 \\ 1 & 0 & 0 & 0 & 0 & 0 & 0 & 0 \\ 0 & 0 & 0 & 1 & 0 & 0 & 0 & 0 \\ 0 & 0 & -1 & 0 & 0 & 0 & 0 & 0 \end{pmatrix}$$, and so on, or in

terms of the inversion matrices

$$\overline{\overline{\alpha}}_x = \begin{pmatrix} 0 & X \\ -X & 0 \end{pmatrix}, \quad \overline{\overline{\alpha}}_y = \begin{pmatrix} 0 & Y \\ -Y & 0 \end{pmatrix}, \quad \overline{\overline{\alpha}}_z = \begin{pmatrix} 0 & Z \\ -Z & 0 \end{pmatrix},$$

$$\overline{\overline{\sigma}}_x = i\begin{pmatrix} X & 0 \\ 0 & X \end{pmatrix}, \quad \overline{\overline{\sigma}}_y = i\begin{pmatrix} Y & 0 \\ 0 & Y \end{pmatrix}, \quad \overline{\overline{\sigma}}_z = \begin{pmatrix} Z & 0 \\ 0 & Z \end{pmatrix}.$$

Multiplication Table for 8-D Matrices

The multiplication table for the $\overline{\overline{\alpha}}$'s and $\overline{\overline{\sigma}}$'s is

	α_x	α_y	α	σ_x	σ_y	σ_z	
α_x	I	-ZI	YI	λ	$-\lambda Z$	λY	
α_y	ZI	I	-XI	λZ	λ	$-\lambda X$	
α_z	-YI	XI	I	$-\lambda Y$	λX	λ	. (7.14)
σ_x	λ	$-\lambda Z$	λY	I	ZI	-YI	
σ_y	λZ	λ	$-\lambda X$	-ZI	I	XI	
σ_z	$-\lambda Y$	λX	λ	YI	-XI	I	

where I is the 8-D identity matrix and

62

$$\lambda = i \begin{pmatrix} 0 & 0 & 0 & 0 & -1 & 0 & 0 & 0 \\ 0 & 0 & 0 & 0 & 0 & -1 & 0 & 0 \\ 0 & 0 & 0 & 0 & 0 & 0 & -1 & 0 \\ 0 & 0 & 0 & 0 & 0 & 0 & 0 & -1 \\ 1 & 0 & 0 & 0 & 0 & 0 & 0 & 0 \\ 0 & 1 & 0 & 0 & 0 & 0 & 0 & 0 \\ 0 & 0 & 1 & 0 & 0 & 0 & 0 & 0 \\ 0 & 0 & 0 & 1 & 0 & 0 & 0 & 0 \end{pmatrix}$$

We could now follow a development in analogy to that presented in Schiff (see Reference 3 - L. I. Schiff, *Quantum Mechanics*, McGraw-Hill, New York (1955), pp. 327 to 337). For example, it is relatively easy to show that

$$\overline{\overline{\beta}}\overline{\overline{\alpha_x}} + \overline{\overline{\alpha_x}}\overline{\overline{\beta}} = 0$$

$$\overline{\overline{\beta}}\overline{\overline{\alpha_y}} + \overline{\overline{\alpha_y}}\overline{\overline{\beta}} = 0 \quad (7.15)$$

$$\overline{\overline{\beta}}\overline{\overline{\alpha_z}} + \overline{\overline{\alpha_z}}\overline{\overline{\beta}} = 0$$

$$\overline{\overline{\beta}}\overline{\overline{\sigma_x}} - \overline{\overline{\sigma_x}}\overline{\overline{\beta}} = 0$$

$$\overline{\overline{\beta}}\overline{\overline{\sigma_y}} - \overline{\overline{\sigma_y}}\overline{\overline{\beta}} = 0 \quad (7.16)$$

$$\overline{\overline{\beta}}\overline{\overline{\sigma_z}} - \overline{\overline{\sigma_z}}\overline{\overline{\beta}} = 0$$

$$\left(\overline{\overline{\alpha}} \square \overline{B} \right)\left(\overline{\overline{\alpha}} \square \overline{C} \right) = \overline{B} \square \overline{C} + i\overline{\overline{\sigma}} \square \left(\overline{B} \times \overline{C} \right) \quad (7.17)$$

where $\overline{B}$ and $\overline{C}$ are arbitrary 3-vectors and, for brevity, we have written the components of $\overline{\overline{\alpha}}$ as $\overline{\overline{\alpha}} = \overline{\overline{\alpha_x}}\hat{i} + \overline{\overline{\alpha_y}}\hat{j} + \overline{\overline{\alpha_z}}\hat{k}$ and the components of $\overline{\overline{\sigma}}$ as $\overline{\overline{\sigma}} = \overline{\overline{\sigma_x}}\hat{i} + \overline{\overline{\sigma_y}}\hat{j} + \overline{\overline{\sigma_z}}\hat{k}$. From equations (7.15) through (7.17), it can be shown that the total angular momentum is

$$\overline{L} + \frac{\hbar\overline{\overline{\sigma}}}{2}$$ and commutes with the Hamiltonian

$$H = -c\overline{\overline{\alpha}}\square\overline{p} - \overline{\overline{\beta}}mc^2 + V \quad (7.18)$$

where V is a spherically symmetric potential. It can also be shown that, if ϕ and $\overline{A}$ are E-M potentials, then the appropriate wave equation for a particle like an electron immersed in an E-M field is

$$\left[\left(\mathcal{E}-e\phi\right)^2-\left(c\overline{p}-e\overline{A}\right)^2-m^2c^4+e\hbar c\overline{\overline{\sigma}}\square\overline{B}+ie\hbar c\overline{\overline{\alpha}}\square\overline{E}\right]\psi=0 \quad (7.19)$$

Equations similar to equations (7.15) through (7.19) can be shown to be true for suitable choices of the 4-D matrices listed in paragraphs A. and B. above. This should lead to solutions , which perhaps give results with correct energy values similar to those in Schiff, but in eight dimensions. However, this development is unlikely to give a lot of new information; consequently, instead of following this development the author prefers to turn to an unconventional approach which differs from that used by Dirac.

The author's approach is to try and use 4-D vectors everywhere in the development; consequently, the first step is to generate a 4-D angular momentum vector. This can be done by operating with the 4-D $\square_1$ matrix for the 4-D position vector on the 4-D momentum vector, $^4\overline{p}=p_4\hat{l}+\overline{p}$, thus

$$^4\overline{L}=\begin{pmatrix} s & x & y & z \\ -x & s & -z & y \\ -y & z & s & x \\ -z & -y & -x & s \end{pmatrix}\begin{pmatrix} p_4 \\ p_x \\ p_y \\ p_z \end{pmatrix}=\left(sp_4+\overline{r}\square\overline{p}\right)\hat{l}+s\overline{p}-\overline{r}p_4+\overline{r}\times\overline{p} \quad (7.20)$$

However, $^4\overline{L}$ does not necessarily commute with the Hamiltonian, $H=-c\overline{\overline{\alpha}}\square\overline{p}-mc^2I+VI$, thus

$$\left\{^4\overline{L},H\right\}=2i\hbar cp_4I .(7.21)$$

The angular momentum commutes with H if the eigenvalue $p_4=0$, of course, but what is the meaning of the term $s\overline{p}$ in equation (7.20) or, for that matter, the operator $p_4=-i\hbar\dfrac{\partial}{\partial s}$? It has become clear now that one should try and understand spin angular momentum better before proceeding through all the labor involved in this approach; consequently, this approach is deferred until sufficient results on spin are obtained.

Chapter VIII. Further Development of Ideas –

Some General Remarks on Vector Forms Used
Equation Section (Next)Equation Section (Next)
Equation Chapter (Next) Section 1Equation Section (Next)
Equation Section (Next)Equation Section (Next)
Equation Section (Next)Equation Section (Next)
Equation Section (Next)Equation Section (Next)

In the Minkowski space that is being employed here, there are two conjugate types of vectors: namely, $iV_4\hat{i} + \bar{V}$ (the first kind) and $U_4\hat{i} + i\bar{U}$ (the second kind), where $\bar{V}$ and $\bar{U}$ are arbitrary spatial 3-vectors. These two types of vectors must be treated as separate and distinct vectors. If these two vectors are squared, we can no longer distinguish which term came from the 4th component and which came from the 3-D space components, thus $-V_4^2 + V^2$ and $U_4^2 - U^2$, unless we know which of the two types of vectors obtain or we have labeled the components in some way, such as a subscript 4 on the time component. One will remember that, in Minkowski space, the magnitude of a vector can be negative as well as positive.

A vector of one kind can be mathematically transformed into one of the other kind by multiplying by $+i$ and inverting the appropriate component (either the 4th component or all three space components at once). Multiplication by i four times in succession returns the vector to its original kind, not surprisingly.

Operation on the first kind of vector with $\square_1$ yields

$$\left(-W_4 V_4 + \overline{W}\overline{\square V}\right)\hat{i} + i\left(-V_4\overline{W} + W_4\overline{V}\right) + \overline{W} \times \overline{V}, \quad (8.1)$$

where $\overline{W} = W_x\hat{i} + W_y\hat{j} + W_z\hat{k}$ is an arbitrary 3-vector and W_4 would be its fourth component, from both of which the $\square_1$ matrices were constructed. Operation with $\square_1^*$ on the first kind of vector yields

$$\left(W_4 V_4 + \overline{W}\overline{\square V}\right)\hat{i} - i\left(V_4\overline{W} + W_4\overline{V}\right) + \overline{W} \times \overline{V}. \quad (8.2)$$

65

On the other hand, operation with $\square_1$ on the second kind of vector, yields

$$i\left(W_4 U_4 + \overline{\overline{W}}\,\overline{\underline{U}}\right)\hat{\imath} - U_4\overline{W} - W_4\overline{U} + i\overline{\overline{W}} \times \overline{U} \,.(8.3)$$

Operation with $\square_1^*$ on the second kind of vector, yields

$$-i\left(W_4 U_4 - \overline{\overline{W}}\,\overline{\underline{U}}\right)\hat{\imath} - U_4\overline{W} - W_4 U + i\overline{\overline{W}} \times \overline{U} \,.(8.4)$$

One must, therefore, use the appropriate operator matrices on each kind of vector to obtain the desired result. The Lorentz transformation, especially, and also the Special Relativity Velocity Addition Rule require not only operation with an appropriate $\square$ matrix, but also that

$$\left(iV'\right)^2 + \left(\overline{V}_4'\right)^2 = \left(iV_4\right)^2 + \left(\overline{V}\right)^2, \qquad (8.5) \text{ where}$$

$$\square_1\begin{pmatrix} iV_4 \\ \overline{V} \end{pmatrix} = \begin{pmatrix} iV_4' \\ \overline{V}' \end{pmatrix} \qquad (8.6)$$

This is accomplished, for example for equations (8.1) and (8.2) by multiplying the $\square_1$ matrix by the factor

$$-\frac{1}{\left(W_4^2 - W^2\right)}, \qquad (8.7) \text{ and for equations (8.3) and (8.4) by}$$

multiplying by the factor

$$\frac{1}{\left(W_4^2 - W^2\right)} \qquad (8.8)$$

The remarkable property of the $\square$ matrices, developed and demonstrated on pages 7 – 9 in Chapter III above, whereby operation with matrices constructed from different (and arbitrary) 4-vectors any number of times on an arbitrary 4-vector yields a 4-vector, the square of which is of the form

$$\overline{w}^2 = \left(a^2 + b^2 + c^2 + d^2\right)\left(e^2 + f^2 + g^2 + h^2\right)(\ldots,\ldots,\ldots,\ldots)\left(s^2 + x^2 + y^2 + z^2\right).(8.9)$$

where a, b, c, d, e, f, g, h, etc. are components of the various vectors. This type of vector lends itself very nicely as we have seen, to construction of the Lorentz transform and of arguments for quantum mechanical wavefunctions.

Now it should be easy to see how to construct a Lorentz transformation that is very generalized. From an arbitrary 4-vector, say $iu_4\hat{l} + \overline{u}$, one gets the generalized Lorentz transform,

$$\mathcal{L}_E = \frac{1}{\sqrt{1 - \left(\dfrac{\overline{u}}{u_4}\right)^2}} \begin{pmatrix} 1 & -i\dfrac{\overline{u}}{u_4} \\ i\dfrac{\overline{u}}{u_4} & 1 \end{pmatrix} \qquad (8.10)$$

And it's conjugate,

$$\mathcal{L}_E^* = \frac{1}{\sqrt{1 - \left(\dfrac{\overline{u}}{u_4}\right)^2}} \begin{pmatrix} 1 & i\dfrac{\overline{u}}{u_4} \\ -i\dfrac{\overline{u}}{u_4} & 1 \end{pmatrix} \qquad (8.11)$$

There is no compelling mathematical reason why one cannot factor the square root radicals differently, so that the additional conjugate transforms,

$$\mathcal{L}_E' \frac{1}{\sqrt{\left(\dfrac{u_4}{u}\right)^2 - 1}} \begin{pmatrix} i & \dfrac{u_4\overline{u}}{u^2} \\ -\dfrac{u_4\overline{u}}{u^2} & i \end{pmatrix} \qquad (8.12) \; and$$

$$\mathcal{L}_E'^* \frac{1}{\sqrt{\left(\dfrac{u_4}{u}\right)^2 - 1}} \begin{pmatrix} -i & \dfrac{u_4\overline{u}}{u^2} \\ -\dfrac{u_4\overline{u}}{u^2} & -i \end{pmatrix} \qquad (8.13) \; \text{are possible. If } \overline{u} \text{ is}$$

taken to be a group velocity, then the first two transforms are transforms with respect to this group velocity, while last two transforms are transforms with respect to a corresponding phase velocity, $\dfrac{u_4\overline{u}}{u^2}$.

Just for fun and for examples, I construct generalized transforms from the energy-momentum 4-vector and the position 4-vector. The energy-momentum 4- vector is,

$$im_0 c^2 \hat{\sigma} = i\mathcal{E}\hat{l} + c\bar{p} \text{ , so that,}$$

$$\mathcal{L}_E = \frac{1}{\sqrt{1-\left(\dfrac{c\bar{p}}{\mathcal{E}}\right)^2}} \begin{pmatrix} 1 & -i\dfrac{c\bar{p}}{\mathcal{E}} \\ i\dfrac{c\bar{p}}{\mathcal{E}} & 1 \end{pmatrix} \qquad (8.14).$$

But $\dfrac{c\bar{p}}{\mathcal{E}} = \dfrac{\gamma m_0 c^2 \bar{\beta}}{\gamma m_0 c^2} = \bar{\beta}$, and thus

$$\mathcal{L}_E = \frac{1}{\sqrt{1-\beta^2}} \begin{pmatrix} 1 & -i\bar{\beta} \\ i\bar{\beta} & 1 \end{pmatrix}.$$

Likewise, the position 4-vector is $ict'\hat{l} + \bar{r}'$ so that

$$\mathcal{L}_E = \frac{1}{\sqrt{1-\left(\dfrac{\bar{r}'}{ct'}\right)^2}} \begin{pmatrix} 1 & -i\dfrac{\bar{r}'}{ct'} \\ i\dfrac{\bar{r}'}{ct'} & 1 \end{pmatrix},$$

However, for constant velocity, $\dfrac{\bar{r}'}{ct'} = \bar{\beta}$, so that again

$$\mathcal{L}_E = \frac{1}{\sqrt{1-\beta^2}} \begin{pmatrix} 1 & -i\bar{\beta} \\ i\bar{\beta} & 1 \end{pmatrix}.$$

The same result is obtained using the electromagnetic vector potential, $i\phi\hat{l} + \phi\bar{\beta}$, for a single charged particle. Apparently the Lorentz transform can be so constructed from any Lorentz invariant 4-vector.

In a more serious vein, I note that the phase velocity is virtually unmentioned and unnoticed in both Special and General relativity and I have shown that there exists, at least mathematically, a Lorentz transform based on the phase velocity. Quantum Mechanics has shown us that waves and particles are aspects of the same thing. The question of the physical applicability of the phase velocity transform does not, therefore, seem to be idle speculation. Likewise a phase velocity gamma, γ_p, should exist.

In the following discussion, except where noted, all velocities will be reduced velocities, i.e. the real velocity divided by the velocity of light in vacuum. Special Relativity requires that all velocities be less than or equal to 1. The general velocity addition rule is framed in the following way:

1.) No velocity component has absolute value greater than 1,

2.) The absolute value of the sum of all velocity components is always less than or equal to 1,

3.) The sum of velocities is itself a vector or pseudovector or combination of the two, and,

4.) If any component of the sum has an absolute value of 1, then the sum must have an absolute value lying between 1 and 0 inclusive.

The addition rule obtained in this work and given by the formula

$$\bar{s} = \frac{\bar{v}_1 + \bar{v}_2 + i\bar{v}_1 \times \bar{v}_2}{1 + \bar{v}_1 \cdot \bar{v}_2} \quad (8.15).$$

This satisfies the above requirements provided that the cross product $\times$ and the dot product $\sqcup$ are suitable cross and dot products for 4 dimensions. The correction factors, $(\bar{v}_1 \times \bar{v}_2)$ and $(\bar{v}_1 \sqcup \bar{v}_2)$ are both less than about 10^{-8} for any velocity less than 18.6 miles per second or about 67,000 mph. This is fast even for a rocket! The correct formula or formulae should be reasonably unique. By this I mean that the formulae for intrinsic velocities of elementary particles may be different from one elementary particle to another and from the formula for normal particle velocities.

In order to determine correct formulae, first, note that the cross product must yield a vector and the dot product must yield a scalar in order for the formula to be consistent. Next, the magnitude of $\bar{S}$ is determined, by squaring,

$$\bar{s}^2 = \left(\frac{\bar{v}_1 + \bar{v}_2 + i\bar{v}_1 \times \bar{v}_2}{1 + \bar{v}_1 \cdot \bar{v}_2} \right)^2 = 1 - \frac{1 - v_1^2 - v_2^2 + (\bar{v}_1 \times \bar{v}_2)^2 + (\bar{v}_1 \cdot \bar{v}_2)^2}{\left(1 + (\bar{v}_1 \cdot \bar{v}_2)\right)^2}$$

(8.16)

$$= 1 - \frac{u}{\left(1 + (\bar{v}_1 \cdot \bar{v}_2)\right)^2}$$

Now if either v_1^2 or $v_2^2 = 1$, then s^2 must equal 1 and this implies that $u = 0$ in that case. If u is zero at $v_1^2 = 1$ or $v_2^2 = 1$, then v_1^2 and v_2^2 are roots of u. Consequently,

69

$$u = u_0\left(1-v_1^2\right)\left(1-v_2^2\right) \quad (8.17)$$

This result in turn implies that

$$v_1^2 v_2^2 = (\bar{v}_1 \times \bar{v}_2)^2 + (\bar{v}_1 \cdot \bar{v}_2)^2. \quad (8.18)$$

Each product of vectors, to make sense, should be proportional to $v_1 v_2$ and the sum of the squares of the proportionality coefficients should equal 1, so that

$$v_1^2 v_2^2 = \left(C_1^2 + C_2^2\right) v_1^2 v_2^2 \;\; and \;\; C_1^2 + C_2^2 = 1.$$

For example, the velocity addition rule is satisfied if one takes either $C_1 = \cos\theta$ and $C_2 = \sin\theta$ or $C_1 = \sin\theta$ and $C_2 = \cos\theta$

The velocity addition rule for parallel velocities is

$$s\hat{p} = \frac{v_1 + v_2}{1 + v_1 v_2}\hat{p} \quad\quad (8.19)$$

And for the sum of 3 velocities

$$s\hat{p} = \frac{v_1 + v_2 + v_3 + v_1 v_2 v_3}{1 + v_1 v_2 + v_1 v_3 + v_2 v_3}\hat{p} \quad (8.20)$$

And for the sum of four velocities

$$s\hat{p} = \frac{v_1 + v_2 + v_3 + v_4 + v_1 v_2 v_3 + v_1 v_2 v_4 + v_1 v_3 v_4 + v_2 v_3 v_4}{1 + v_1 v_2 + v_1 v_3 + v_1 v_4 + v_2 v_3 + v_2 v_4 + v_3 v_4 + v_1 v_2 v_3 v_4}\hat{p} \, (8.21)$$

And so on. For components that are equal in magnitude, the sum of n components is

$$s\hat{p} = \frac{(1+v)^n - (1-v)^n}{(1+v)^n + (1-v)^n}\hat{p} \quad\quad (8.22)$$

Which for n very, very large and $v \le 1$, becomes

$$s\hat{p} = \tanh(nv)\hat{p} \quad\quad (8.23)$$

One should note that

$$\tanh(z_1 + z_2) = \frac{\tanh(z_1) + \tanh(z_2)}{1 + \tanh(z_1)\tanh(z_2)}, (8.24)$$

which is formally the same as equation (8.19).

The sum for two mutually perpendicular components is

$$\bar{s} = \bar{v}_1 + \bar{v}_2 + i\bar{v}_1 \times \bar{v}_2 \quad (8.25) \; or$$

$$\bar{s} = v_1 \hat{i} + v_2 \hat{j} + i v_1 v_2 \hat{k} \quad (8.26)$$

The sum for three mutually perpendicular components is

70

$$\bar{s} = \frac{\left(v_1 + iv_2v_3\right)\hat{i} + \left(v_2 - iv_1v_3\right)\hat{j} + \left(v_3 + iv_1v_2\right)\hat{k}}{1 + iv_1v_2v_3}, \quad (8.27) \quad \text{if} \quad \text{one}$$

uses the multiplication tables listed in the next chapter for the cross products of the unit vectors. Addition of four mutually perpendicular velocities is a special case and will be discussed below. To go to five and more mutually perpendicular velocities, it will be necessary to introduce 8-dimensional or 16-dimensional multiplication tables, for instance.

Chapter IX. Further Development of Ideas –
The Special Relativity Velocity Addition Rule

Equation Section (Next)

The reader will recall that in Chapter V, the author constructed a generalized Lorentz transformation based on the □ matrix, see equations (5.2), (5.16), and (5.17) which are reproduced below for convenience in slightly modified and simplified form.

$$\pmb{\mathcal{L}}_s = \gamma_\beta \begin{bmatrix} 1 & -i\bar{\beta} \\ i\bar{\beta} & 1 \end{bmatrix} \quad (5.2)$$

On occasion the infinitesimal Lorentz transform will be used. It is constructed from the author's Lorentz transform simply by setting $\gamma = 1$. Note that the infinitesimal Lorentz transform is not the Galilean transform. The former transforms the 4-vector, $ict\hat{l} + \bar{r}$ to $i\left(ct - \bar{\beta}\Box\bar{r}\right)\hat{l} + \bar{r} - \bar{\beta}ct - i\bar{\beta} \times \bar{r}$ while the latter transforms it to $ict\hat{l} + \bar{r} - \bar{\beta}ct$ only (disregarding the fact that the Galilean transform is also only three dimensional). That is, the Galilean transform requires not only the velocity, β, to be small, but also the retardation time, $t_r = r/c$, to be small.

The velocity addition rule was obtained by applying $\pmb{\mathcal{L}}_s$ twice to the position 4-vector and is reproduced here for convenience:

$$\bar{u} = \frac{\bar{\alpha} + \bar{\beta} + i(\bar{\alpha} \times \bar{\beta})}{1 + \bar{\alpha}\Box\bar{\beta}} \quad (5.15) \ and$$

$$\gamma_u = \gamma_\alpha \gamma_\beta (1 + \bar{\alpha}\Box\bar{\beta}). \quad (5.16)$$

(The reader should keep in mind that it is the magnitude of the sum, $\bar{u}$, that is important physically, so that the order of addition will not necessarily be important in the physics.) Equations (5.15) and (5.16) will hereafter be referred to as the Special Relativity Velocity Addition Rule, SRVAR for short. The SRVAR is very powerful because it allows one to add velocity components that are at any angle to each other including right angles and therefore, permits addition of velocity vectors; in contrast to the conventional addition law which is only good if the velocity components are parallel.

From the basic principle of special relativity (the constancy of c in vacuum and its role as the upper limit on any velocity), the SRVAR should apply in every case. However, for two cases it is not immediately clear whether the SRVAR applies without modification. The first is addition of a spatial velocity to the fourth component of velocity or vice versa and the second is addition of continuously changing velocities. First I will take up the addition of a more or less arbitrary spatial velocity 3-vector, say $\bar{a}$, to a similarly arbitrary time or 4-component of velocity, say $id\hat{l}$. The presence of a cross product in the SRVAR makes it necessary to determine the multiplication table for cross products of all unit vectors in any 4-vector. From experience in determining the cross products of multidimensional vectors (see Reference 1 - "*A Method for Constructing the Vector or Cross Product for Multi-dimensional Vectors*", Clifford E. Morgan, U.S. Army Science and Technology Center, Mainz-Kastel, Germany, Copyright August 1997), the form of the $\square$ matrix yields directly the correct multiplication table, shown below:

$$
\begin{array}{c|cccc}
 & \hat{l} & \hat{i} & \hat{j} & \hat{k} \\
\hat{l} & 0 & \hat{i} & \hat{j} & \hat{k} \\
\hat{i} & -\hat{i} & 0 & \hat{k} & -\hat{j} \\
\hat{j} & -\hat{j} & -\hat{k} & 0 & \hat{i} \\
\hat{k} & -\hat{k} & \hat{j} & -\hat{i} & 0
\end{array} \quad (9.1)
$$

That is the form of the table for a cross product should be completely asymmetric. With use of this multiplication table, the SRVAR yields the following sum:

$$\bar{s}_f = id\hat{l} \oplus \bar{a} = id\hat{l} + \bar{a} + i(id\hat{l}) \times (a\hat{j}) = id\hat{l} + (1-d)\bar{a} \qquad (9.2)$$

From now on the symbol $\oplus$ will be used to indicate addition by SRVAR. Now adding in the reverse order, one gets the sum,

$$\bar{s}_r = \bar{a} \oplus id\hat{l} = id\hat{l} + (1+d)\bar{a} \qquad (9.3)$$

It is possible to get the same results with a transform which is similar to a Lorentz transform, except that two different matrices are needed. Thus, the forward matrix is,

$$\bar{s}_f = \gamma_1 \begin{bmatrix} 1 & i\dfrac{(1-d)\bar{a}}{d} \\ -i\dfrac{(1-d)\bar{a}}{d} & 1 \end{bmatrix} \begin{bmatrix} i\hat{l} \\ 0 \end{bmatrix} = \begin{bmatrix} i\gamma_1\hat{l} \\ \gamma_1\dfrac{(1-d)\bar{a}}{d} \end{bmatrix} \quad (9.4)$$

And if $\gamma_1 = d$, the resultant vector in equation (9.4) becomes $id\hat{l} + (1-d)\bar{a}$, as required. The reverse matrix is,

$$\bar{s}_r = \gamma_2 \begin{bmatrix} 1 & i\dfrac{d\bar{a}}{(1+d)a^2} \\ -i\dfrac{d\bar{a}}{(1+d)a^2} & 1 \end{bmatrix} \begin{bmatrix} 0 \\ \bar{a} \end{bmatrix} = \begin{bmatrix} \dfrac{i\gamma_2 d}{(1+d)}\hat{l} \\ \gamma_2\bar{a} \end{bmatrix} \quad (9.5)$$

And if $\gamma_2 = (1+d)$, the vector $id\hat{l} + (1+d)\bar{a}$ ensues as required. There are, of course, conditions, both physical and mathematical, on the magnitudes of d and (1-d)a. A first condition is that they must lie between 0 and 1. A second condition is that the 4-vectors must be Lorentz invariant; or, relaxing this somewhat, they should have an invariant magnitude equal to the rest magnitude. A third possible requirement is the rest mass of a particle should be invariant to any changes of velocity, i.e. either $\bar{a}$ is always constant or $\bar{s}_f^2$ and $\bar{s}_r^2$ are independent of any velocity $\bar{a}$. A fourth requirement might be that $\bar{s}_f = \bar{s}_r$, i.e. addition in the forward mathematical direction should give the same results as addition in the reverse mathematical direction (this depends, of course, on the physics of the process, e.g. creation and destruction of particles may not be exactly time reversible). The simplest way to satisfy the invariance requirement is to set $d = (1-d)$ in. equation (9.2) and $-d = (1+d)$ in equation (9.3). This yields $d = \dfrac{1}{2}$ and $d = -\dfrac{1}{2}$ respectively. These values of d give $\gamma_1 = \dfrac{1}{2}$ and $\gamma_2 = \dfrac{1}{2}$. Equations (9.4) and (9.5) become

$$\bar{s}_f = \frac{1}{2}\begin{bmatrix} 1 & i\bar{a} \\ -i\bar{a} & 1 \end{bmatrix}\begin{bmatrix} i\hat{l} \\ 0 \end{bmatrix} = \begin{bmatrix} \dfrac{i}{2}\hat{l} \\[2mm] \dfrac{\bar{a}}{2} \end{bmatrix} \qquad \textit{(9.6) and}$$

$$\bar{s}_r = \frac{1}{2}\begin{bmatrix} 1 & -i\dfrac{\bar{a}}{a^2} \\[2mm] i\dfrac{\bar{a}}{a^2} & 1 \end{bmatrix}\begin{bmatrix} 0 \\ \bar{a} \end{bmatrix} = \begin{bmatrix} -\dfrac{i}{2}\hat{l} \\[2mm] \dfrac{\bar{a}}{2} \end{bmatrix} \qquad \textit{(9.7)}$$

This means that the time component is not necessarily reversible, if particle creation or destruction occurs. (The above result is, of course consistent with the multiplication table that I chose on the preceding page.)

One takes the masses and energies of the particles to be either positive or negative. In the Lorentz transforms given by equations (9.4) and (9.5) then, the gammas must be positive. It is emphasized that $\frac{1}{2}$ is the only value the gammas can have in this simple case, i.e.

$$\gamma_1 = \gamma_2 = \frac{1}{2} \text{ , so that}$$

$$\bar{s} = i\frac{1}{2}\hat{l} + \frac{1}{2}\bar{a} \qquad (9.8)$$

Note that if $|\bar{a}|$ is set equal to $i\sqrt{3}$, then $s^2 = -\frac{1}{4} - \frac{3}{4} = -1$, so that the rest mass is quantized correctly, i.e. the rest mass equals m_0, and if $|\bar{a}|$ is set equal to 1, then $s^2 = -\frac{1}{4} + \frac{1}{4} = 0$, so that the rest mass is quantized to 0; but, in both cases, the effective gamma is still equal to $\frac{1}{2}$.

These results open the door to a reasonable model for spin (see Chapter X). Nevertheless, however, I will consider in the following a perhaps better and more interesting method of development which leads to a greater diversity of results.

That method is to begin by squaring equation (9.2)and equation (9.3) to get,

$$s_f^2 = -d^2 + (1-d)^2 a^2 \qquad (9.9) \text{ and}$$

$$s_r^2 = -d^2 + (1+d)^2 a^2 \qquad (9.10).$$

Now then, if one sets

$$-d^2 = (1-d)^2 \text{ in equation (9.9) and}$$

$$-d^2 = (1+d)^2 \text{ in equation (9.10), one gets}$$

$$d = \frac{1}{2} \pm \frac{i}{2} \text{ and}$$

$d = -\frac{1}{2} \pm \frac{i}{2}$ respectively. Substituting these results in the 4-vectors, equations (9.2) and (9.3), and taking the upper sign only, one obtains

$$\overline{s}_f = i\left(\frac{1}{2} + \frac{i}{2}\right)\hat{i} + \left(\frac{1}{2} - \frac{i}{2}\right)\overline{a} = \left(\frac{i}{2}\hat{i} + \frac{1}{2}\overline{a}\right) - \left(\frac{1}{2}\hat{i} + \frac{i}{2}\overline{a}\right) \text{(9.11) and}$$

$$\overline{s}_r = i\left(-\frac{1}{2} + \frac{i}{2}\right)\hat{i} + \left(\frac{1}{2} + \frac{i}{2}\right)\overline{a} = \left(-\frac{i}{2}\hat{i} + \frac{1}{2}\overline{a}\right) + \left(-\frac{1}{2}\hat{i} + \frac{i}{2}\overline{a}\right) \text{(9.12)}$$

Further,

$$s_f^2 = -\frac{i}{2}\left(1 + a^2\right) \qquad (9.13) \text{ and}$$

$$s_r^2 = \frac{i}{2}\left(1 + a^2\right) \text{(9.14)}$$

Rest mass quantization is preserved in both equations (9.13) and (9.14) if $a^2 = 1$. Now taking the lower signs in the equations for d and substituting for d in equations (9.2) and (9.3), one obtains the 4-vectors

$$\overline{s}_f = i\left(\frac{1}{2} - \frac{i}{2}\right)\hat{i} + \left(\frac{1}{2} + \frac{i}{2}\right)\overline{a} = \left(\frac{i}{2}\hat{i} + \frac{1}{2}\overline{a}\right) + \left(\frac{1}{2}\hat{i} + \frac{i}{2}\overline{a}\right) \qquad (9.15)$$

$$\overline{s}_r = i\left(-\frac{1}{2} - \frac{i}{2}\right)\hat{i} + \left(\frac{1}{2} - \frac{i}{2}\right)\overline{a} = \left(-\frac{i}{2}\hat{i} + \frac{1}{2}\overline{a}\right) + \left(\frac{1}{2}\hat{i} - \frac{i}{2}\overline{a}\right) \qquad (9.16)$$

Hence,

$$s_f^2 = \frac{i}{2}\left(1 + a^2\right) \qquad (9.17)$$

76

$$s_r^2 = -\frac{i}{2}\left(1 + a^2\right) \qquad (9.18)$$

In both cases, if $a^2 = 1$, then the rest mass is properly quantized.

In order to show that the above results can yield the normal addition of angular momenta, equation (9.2) is considered again:

$$\bar{s} = id\hat{l} + (1 - d)\bar{a} \ .$$

Adding $(n-1)\bar{a}$ to this equation by the ordinary vector addition (one is now adding particles, not velocities), one gets

$$\bar{s}' = id\hat{l} + (1 - d)\bar{a} + (n-1)\bar{a} = id\hat{l} + (n - d)\bar{a} \quad \text{where} \quad n \text{ is a}$$

positive integer. Applying the squaring procedure used above and setting coefficients of the velocities equal one obtains

$$-d^2 = (n - d)^2 \text{ then,}$$

$$d = \frac{n}{2}(1 \pm i)$$

Substituting this result into the equation for $\bar{s}'$ above, one has

$$\bar{s}' = \left(i\frac{n}{2}(1 \pm i)\hat{l} + \frac{n}{2}(1 \mp i)\bar{a} \right)$$

One can easily see that if $\bar{a}$ is interpreted as the velocity of the quantized intrinsic angular momentum of n identical elementary particles, then the total spin angular momentum of the particles is given by the equation for $\bar{s}'$ immediately above. Note that

$$s'^2 = \mp i\frac{n^2}{2}\left(1 + a^2\right) \ , \text{ so that if } a^2 = 1 \text{ , then } s'^2 = in^2 \text{ and the rest}$$

mass remains properly quantized to m_0 .

Now I consider the addition of two velocities $id\hat{l}$ and $ih\hat{l}$ by the SRVAR which we have, so far, neglected. The sum, $\bar{s}$, equals

$$\bar{s} = i\left(\frac{d + h}{1 + dh}\right)\hat{l} \qquad (9.19).$$

Is it possible to construct a Lorentz transform which gives this result when operating on the vector, $id\hat{l}$? The answer, of course, is yes and the transform matrix is

$$\mathcal{L}_1 = \gamma \begin{bmatrix} h & 0 \\ 0 & h \end{bmatrix} \ (9.20).$$

So that

$$\mathcal{L}_d \begin{bmatrix} id\hat{\imath} \\ 0 \end{bmatrix} = \gamma \begin{bmatrix} h & 0 \\ 0 & h \end{bmatrix} \begin{bmatrix} id\hat{\imath} \\ 0 \end{bmatrix} = i\gamma hd\hat{\imath} \quad (9.21)$$

Hence, if $\bar{s} = i\left(\dfrac{d+h}{1+dh}\right)\hat{\imath} = i\gamma hd\hat{\imath}$, post hoc; ergo propter hoc,

$$\gamma = \frac{1}{hd}\left(\frac{d+h}{1+dh}\right) \quad (9.22).$$

Adding $ih\hat{\imath}$ to $id\hat{\imath}$ corresponds to adding mass, or energy to a particle of rest mass, $m_0 d$.

I continue here with some more discussion of both the SRVAR and the Lorentz transform, $\mathcal{L}_1$. It is interesting to see that the differential □ operator, D, can be made to look like the operator $\mathcal{L}_1$ with the aid of a wavefunction, ψ. One simply operates with a slightly modified D operator on the position 4-vector multiplied by ψ, thus

$$D_L \begin{bmatrix} ict\psi\hat{\imath} \\ \bar{r}\psi \end{bmatrix} = \begin{bmatrix} \dfrac{n}{c}\dfrac{\partial}{\partial t} & -i\nabla \\ i\nabla & \dfrac{n}{c}\dfrac{\partial}{\partial t} \end{bmatrix} \begin{bmatrix} ict\psi\hat{\imath} \\ \bar{r}\psi \end{bmatrix} = \begin{bmatrix} i\left(n\psi + nt\dfrac{\partial\psi}{\partial t} - \nabla\Box\bar{r}\psi - \bar{r}\Box\nabla\psi\right)\hat{\imath} \\ -ct\nabla\psi + \dfrac{n}{c}\dfrac{\partial\bar{r}}{\partial t}\psi + \dfrac{n\bar{r}}{c}\dfrac{\partial\psi}{\partial t} - i\bar{r}\times\nabla\psi \end{bmatrix} \quad (9.23)$$

Where n is the number of space dimensions of $\bar{r}$ and

$$\psi = \psi_0 \exp\left[\gamma k\left(\bar{\beta}\Box\bar{r} + \frac{ct}{n}\right)\right].$$

If $\bar{r}$ is the rest frame position vector, then $\dfrac{\partial\bar{r}}{\partial t} = 0$, and one has,

$$D_L \begin{bmatrix} ict\psi\hat{\imath} \\ \bar{r}\psi \end{bmatrix} = \gamma k\left[i\left(ct - \bar{\beta}\Box\bar{r}\right)\psi\hat{\imath} + \left(\bar{r} - \bar{\beta}ct\right)\psi + i\left(\bar{\beta}\times\bar{r}\right)\psi\right] \quad (9.24)$$

Now if $k_1 = \gamma k$, the resultant vector in equation (9.24) looks exactly like the $\mathcal{L}_1$ transform of the position vector times $k_0\psi$. That is,

$$D_L \begin{bmatrix} ict\psi\hat{\imath} \\ \bar{r}\psi \end{bmatrix} = k\left[\gamma(ct - \bar{\beta}\Box\bar{r})\hat{\imath} + \gamma(\bar{r} - \bar{\beta}ct) - i\gamma\bar{\beta}\times\bar{r} \right]\psi = k\psi\mathcal{L}_1 \begin{bmatrix} ict\hat{\imath} \\ \bar{r} \end{bmatrix}$$

$$(9.25)$$

Note that the wave function, ψ, has only the phase velocity,

$\bar{v}_p = \dfrac{c\bar{\beta}}{n\beta^2}$, and no group velocity because $\dfrac{\partial \bar{r}}{\partial t} = 0$, and further, the

wave function is a plane wave extending over all space.

Further, if the wavefunction in equation (9.24) is replaced by

$\psi = \psi_0 \exp\left[\gamma_\alpha\left(\bar{\alpha}\Box\bar{r} + \dfrac{ct}{n} \right) \right]$ and one operates one more time with

the D_L operator and setting the group velocity again to $\dfrac{\partial \bar{r}}{\partial t} = 0$ and

the constant, $k_2 = \gamma_\alpha$, one gets the transformed vector,

$$ik\gamma_\alpha\gamma_\beta\left[ct\left(1 + \bar{\alpha}\Box\bar{\beta}\right) - \left(\bar{\alpha} + \bar{\beta} + i\bar{\alpha}\times\bar{\beta}\right)\Box\bar{r} \right]\psi\hat{\imath} +$$

$$k\gamma_\alpha\gamma_\beta\left[\bar{r}\left(1 + \bar{\alpha}\Box\bar{\beta}\right) - \left(\bar{\alpha} + \bar{\beta} + i\left(\bar{\alpha}\times\bar{\beta}\right)\right)ct + i\left(\bar{\alpha} + \bar{\beta} + i\left(\bar{\alpha}\times\bar{\beta}\right)\right)\times\bar{r} \right]\psi$$

Provided, $\nabla\Box\left(\bar{\beta}\times\bar{r}\right) = 0$, and $\nabla\times\left(\bar{\beta}\times\bar{r}\right) = (n-1)\bar{\beta}$. Now one

can see that the D operator is closely related to the Lorentz transform.

Hence, gamma times the differential operator operating on a 4-vector times a suitable wavefunction may, perhaps, yield the same result as a constant times the wavefunction times the Lorentz transform matrix operating on a corresponding 4-vector for certain special 4-vectors. This is a very interesting possibility; so interesting that we try it for the energy-momentum vector. Because γ times the D matrix operating on the energy-momentum 4-vector yields a force 4-vector, we expect this to be, in the same way as above, equal to the Lorentz transform of the force 4-vector. Consequently,

$$\gamma D_L \begin{bmatrix} i\mathcal{E}\psi\hat{\imath} \\ c\bar{p}\psi \end{bmatrix} = -k\psi\,\pounds \begin{bmatrix} iF_4\hat{\imath} \\ \bar{F} \end{bmatrix}$$

$$\begin{bmatrix} \dfrac{1}{c}\dfrac{\partial}{\partial t} & -i\nabla \\[2mm] i\nabla & \dfrac{1}{c}\dfrac{\partial}{\partial t} \end{bmatrix} \begin{bmatrix} i\mathcal{E}\psi\hat{\imath} \\ c\bar{p}\psi \end{bmatrix} = -k\psi \begin{bmatrix} 1 & -i\bar{\beta} \\ i\bar{\beta} & 1 \end{bmatrix} \begin{bmatrix} iF_4\hat{\imath} \\ \bar{F} \end{bmatrix}$$

$$i\left[\left(\frac{1}{c}\frac{\partial\mathcal{E}}{\partial t}-c\nabla\square\bar{p}\right)\psi+\frac{i\mathcal{E}}{c}\frac{\partial\psi}{\partial t}-ic\bar{p}\square\nabla\psi\right]\hat{\imath}+\left(-\nabla\mathcal{E}+\frac{\partial\bar{p}}{\partial t}\right)\psi-\mathcal{E}\nabla\psi+\bar{p}\frac{\partial\psi}{\partial t}$$

$$-ic\left(\nabla\times\bar{p}\right)\psi-ic\bar{p}\times\nabla\psi=-k\psi\left[i\left(F_4-\bar{\beta}\square\bar{F}\right)\hat{\imath}+\bar{F}-\bar{\beta}F_4-i\bar{\beta}\times\bar{F}\right]$$

(9.26)

Taking $\nabla\square\bar{p}=0$ and $\nabla\times\bar{p}=0$ and setting corresponding components equal, one gets

$$\frac{1}{c}\frac{\partial\mathcal{E}}{\partial t}\psi+\frac{\mathcal{E}}{c}\frac{\partial\psi}{\partial t}+c\bar{p}\square\nabla\psi=-F_4\psi+\bar{\beta}\square\bar{F}\psi$$

(9.27)

$$\nabla\mathcal{E}\psi+\mathcal{E}\nabla\psi+\frac{\partial\bar{p}}{\partial t}\psi+\bar{p}\frac{\partial\psi}{\partial t}-ic\bar{p}\times\nabla\psi=\bar{F}-\bar{\beta}F_4-i\bar{\beta}\times\bar{F}$$

Clearly, the rate of change of the energy is the work done by the force on the particle and the force on the particle is the time rate of change of its momentum plus the gradient of any potentials, thus,

$$\frac{\partial\mathcal{E}}{\partial t}=c\bar{\beta}\square\bar{F}$$

(9.28)

$$\bar{F}=\frac{\partial\bar{p}}{\partial t}+\nabla\mathcal{E}$$

This leaves the equations:

$$F_4\psi=-\frac{\mathcal{E}}{c}\frac{\partial\psi}{\partial t}-cp\square\nabla\psi$$

(9.29)

$$\mathcal{E}\nabla\psi+\bar{p}\frac{\partial\psi}{\partial t}-ic\bar{p}\times\nabla\psi=-\bar{\beta}F_4\psi-i\bar{\beta}\times\bar{F}\psi$$

Now, if the wavefunction, ψ, is chosen such that $i\hbar\dfrac{\partial\psi}{\partial t} = \mathcal{E}\psi$ and $-i\hbar\nabla\psi = \overline{p}\psi$, one gets:

$$F_4 = \frac{i}{\hbar c}\left(2\mathcal{E}^2 - m_0^2 c^4\right)$$

$$(9.30)$$

$$\overline{\beta}F_4 = -\frac{2i}{\hbar c}\mathcal{E}^2\overline{\beta}$$

Which lead to the equation:

$$-2\frac{i}{\hbar c}\mathcal{E}^2 = \frac{i}{\hbar c}\left(2\mathcal{E}^2 - m_0^2 c^4\right)$$

And the result that $\mathcal{E} = \dfrac{m_0 c^2}{2}$ and, consequently,

$$F_4 = -\frac{ik_0 m_0 c^2}{2} . \qquad (9.31)$$

Clearly gamma is equal to $\dfrac{1}{2}$. The 4-component of the force, F_4, could be the force that holds the spinning particle in its spin orbit. It is the right order of magnitude to hold the spinning particle in a circular orbit of radius $r_0 = \dfrac{1}{k_0}$ as would be required. Hence, these 4-vectors and their treatment above may apply to the case of particle spin.

Next a generalization is proposed and outlined in this, the last part of the chapter. This generalization is in the same spirit as generalization of the classical angular momentum to include half-integral spin in quantum mechanics. In that case, the classical definition of angular momentum, namely, $\overline{L} = \overline{r} \times \overline{p}$ is supplanted by the commutation relations,

$$L_x L_y - L_y L_x = i\hbar L_z$$

$$L_z L_x - L_x L_z = i\hbar L_y \qquad (9.32)$$

$$L_y L_z - L_z L_y = i\hbar L_x$$

The 4-dimensional matrices,

$$\alpha_j = \begin{bmatrix} \sigma_j & 0 \\ 0 & \sigma_j \end{bmatrix} \quad \alpha_0 = \begin{bmatrix} I & 0 \\ 0 & I \end{bmatrix}$$

$$\beta_j = \begin{bmatrix} 0 & \sigma_j \\ \sigma_j & 0 \end{bmatrix} \quad \beta_0 = \begin{bmatrix} 0 & I \\ I & 0 \end{bmatrix}$$

(9.33) are introduced,

$$\gamma_j = \begin{bmatrix} \sigma_j & 0 \\ 0 & -\sigma_j \end{bmatrix} \quad \gamma_0 = \begin{bmatrix} I & 0 \\ 0 & -I \end{bmatrix}$$

$$\delta_j = \begin{bmatrix} 0 & \sigma_j \\ -\sigma_j & 0 \end{bmatrix} \quad \delta_0 = \begin{bmatrix} 0 & I \\ -I & 0 \end{bmatrix}$$

where the σ_j are the Pauli spin matrices,

$$\sigma_x = \begin{bmatrix} 0 & 1 \\ 1 & 0 \end{bmatrix} \quad \sigma_y = \begin{bmatrix} 0 & -i \\ i & 0 \end{bmatrix} \quad \sigma_z = \begin{bmatrix} 1 & 0 \\ 0 & -1 \end{bmatrix} \text{, and}$$

$$I = \begin{bmatrix} 1 & 0 \\ 0 & 1 \end{bmatrix}$$

The new 4-dimensional matrices can be used for the basis or unit vectors of the spin angular momentum components, of course. For possible future use, the multiplication tables for these basis vectors are recorded in the following multiplication tables:

Multiplication Table I

	α_0	α_x	α_y	α_z	β_0	β_x	β_y	β_z
α_0	α_0	α_x	α_y	α_z	β_0	β_x	β_y	β_z
α_x	α_x	α_0	$i\alpha_z$	$-i\alpha_y$	β_x	β_0	$i\beta_z$	$-i\beta_y$
α_y	α_y	$-i\alpha_z$	α_0	$i\alpha_x$	β_y	$-i\beta_z$	β_0	$i\beta_x$
α_z	α_z	$i\alpha_y$	$-i\alpha_x$	α_0	β_z	$i\beta_y$	$-i\beta_x$	β_0
β_0	β_0	β_x	β_y	β_z	α_0	α_x	α_y	α_z
β_x	β_x	β_0	$i\beta_z$	$-i\beta_y$	α_x	α_0	$i\alpha_z$	$-i\alpha_y$
β_y	β_y	$-i\beta_z$	β_0	$i\beta_x$	α_y	$-i\alpha_z$	α_0	$i\alpha_x$
β_z	β_z	$i\beta_y$	$-i\beta_x$	β_0	α_z	$i\alpha_y$	$-i\alpha_x$	α_0

(9.34)

Multiplication Table II

	γ_0	γ_x	γ_y	γ_z	δ_0	δ_x	δ_y	δ_z	
α_0	γ_0	γ_x	γ_y	γ_z	δ_0	δ_x	δ_y	δ_z	
α_x	γ_x	γ_0	$i\gamma_z$	$-i\gamma_y$	δ_x	δ_0	$i\delta_z$	$-i\delta_y$	
α_y	γ_y	$-i\gamma_z$	γ_0	$i\gamma_x$	δ_y	$-i\delta_z$	δ_0	$i\delta_x$	
α_z	γ_z	$i\gamma_y$	$-i\gamma_x$	γ_0	δ_z	$i\delta_y$	$-i\delta_x$	δ_0	*(9.35)*
β_0	$-\delta_0$	$-\delta_x$	$-\delta_y$	$-\delta_z$	$-\gamma_0$	$-\gamma_x$	$-\gamma_y$	$-\gamma_z$	
β_x	$-\delta_x$	δ_0	$-i\delta_z$	$i\delta_y$	$-\gamma_x$	$-\gamma_0$	$-i\gamma_z$	$i\gamma_y$	
β_y	$-\delta_y$	$i\delta_z$	β_0	$-i\delta_x$	$-\gamma_y$	$i\gamma_z$	$-\gamma_0$	$-i\gamma_x$	
β_z	$-\delta_z$	$-i\delta_y$	$i\delta_x$	δ_0	$-\gamma_z$	$-i\gamma_y$	$i\gamma_x$	$-\gamma_0$	

Multiplication Table III

	α_0	α_x	α_y	α_z	β_0	β_x	β_y	β_z	
γ_0	γ_0	γ_x	γ_y	γ_z	δ_0	δ_x	δ_y	δ_z	
γ_x	γ_x	γ_0	$i\gamma_z$	$-i\gamma_y$	δ_x	δ_0	$i\delta_z$	$-i\delta_y$	
γ_y	γ_y	$-i\gamma_z$	γ_0	$i\gamma_x$	δ_y	$-i\delta_z$	δ_0	$i\delta_x$	
γ_z	γ_z	$i\gamma_y$	$-i\gamma_x$	γ_0	δ_z	$i\delta_y$	$-i\delta_x$	δ_0	*(9.36)*
δ_0	δ_0	δ_x	δ_y	δ_z	γ_0	γ_x	γ_y	γ_z	
δ_x	δ_x	δ_0	$i\delta_z$	$-i\delta_y$	γ_x	γ_0	$i\gamma_z$	$-i\gamma_y$	
δ_y	δ_y	$-i\delta_z$	β_0	$i\delta_x$	γ_y	$-i\gamma_z$	γ_0	$i\gamma_x$	
δ_z	δ_z	$i\delta_y$	$-i\delta_x$	δ_0	γ_z	$i\gamma_y$	$-i\gamma_x$	γ_0	

Multiplication Table IV

	γ_0	γ_x	γ_y	γ_z	δ_0	δ_x	δ_y	δ_z	
γ_0	α_0	α_x	α_y	α_z	β_0	β_x	β_y	β_z	
γ_x	α_x	α_0	$i\alpha_z$	$-i\alpha_y$	β_x	β_0	$i\beta_z$	$-i\beta_y$	
γ_y	α_y	$-i\alpha_z$	α_0	$i\alpha_x$	β_y	$-i\beta_z$	β_0	$i\beta_x$	
γ_z	α_z	$i\alpha_y$	$-i\alpha_x$	α_0	β_z	$i\beta_y$	$-i\beta_x$	β_0	(9.37)
δ_0	$-\beta_0$	$-\beta_x$	$-\beta_y$	$-\beta_z$	$-\alpha_0$	$-\alpha_x$	$-\alpha_y$	$-\alpha_z$	
δ_x	β_x	β_0	$-i\beta_z$	$i\beta_y$	$-\alpha_x$	$-\alpha_0$	$-i\alpha_z$	$i\alpha_y$	
δ_y	β_y	$i\beta_z$	β_0	$-i\beta_x$	$-\alpha_y$	$i\alpha_z$	$-\alpha_0$	$-i\alpha_x$	
δ_z	β_z	$-i\beta_y$	$i\beta_x$	β_0	$-\alpha_z$	$-i\alpha_y$	$i\alpha_x$	$-\alpha_0$	

These four Tables can be fitted together into one 16×16 Table as shown below:

$$\begin{bmatrix} Table\ I & Table\ II \\ Table\ III & Table\ IV \end{bmatrix}$$ resulting in a Grand Multiplication Table. The Grand Multiplication Table is potentially useful enough in further development that it is listed here for reference.

Another step farther taken down this road would be to introduce the additional matrices:

$$\varepsilon_j = \begin{bmatrix} \sigma_j & 0 \\ 0 & i\sigma_j \end{bmatrix} \quad \varepsilon_0 = \begin{bmatrix} I & 0 \\ 0 & iI \end{bmatrix}$$

$$\varsigma_j = \begin{bmatrix} 0 & \sigma_j \\ i\sigma_j & 0 \end{bmatrix} \quad \varsigma_j = \begin{bmatrix} 0 & I \\ iI & 0 \end{bmatrix}$$

$$\eta_j = \begin{bmatrix} \sigma_j & 0 \\ 0 & -i\sigma_j \end{bmatrix} \quad \eta_0 = \begin{bmatrix} I & 0 \\ 0 & -iI \end{bmatrix}$$

$$\xi_j = \begin{bmatrix} 0 & \sigma_j \\ -i\sigma_j & 0 \end{bmatrix} \quad \xi_j = \begin{bmatrix} 0 & I \\ -iI & 0 \end{bmatrix}$$

(9.38), leading to a colossal 32×32 Multiplication Table. The final step would be to introduce the following multiplication table:

Final Multiplication Table

	d	a	b	c	h	e	f	g	
d	d	a	b	c	h	e	f	g	
a	$-a$	d	$-c$	b	$-e$	h	$-g$	f	
b	$-b$	c	d	$-a$	$-f$	g	h	$-e$	
c	$-c$	$-b$	a	d	$-g$	$-f$	e	h	(9.39)
h	$-h$	ie	if	ig	d	$-a$	$-b$	$-c$	
e	$-ie$	$-h$	$-ig$	if	a	d	c	$-b$	
f	$-if$	ig	$-h$	$-ie$	b	$-c$	d	a	
g	$-ig$	$-if$	ie	$-h$	c	b	$-a$	d	

This multiplication table generates what I will call lop-sided cross products. These are cross products which anticommute with a vengeance, i.e., for example, $\bar{a} \times \bar{f} = -\bar{g}$ and $\bar{f} \times \bar{a} = i\bar{g}$ so that $\left(\bar{a} \times \bar{f}\right)^2 = \left(\bar{g}\right)^2$ and $\left(\bar{f} \times \bar{a}\right)^2 = -\left(\bar{g}\right)^2$ etc. Thus these products, even when squared, do not commute. They "remember" the direction of multiplication. With this last step, a degree of complexity is reached which is bad enough to cause one's head to spin. Nevertheless, these objects may actually be useful.

Chapter X. Further Development of Ideas – Spin and a Spin Model

Equation Section (Next)

All results pertaining to intrinsic momentum of a particle are collected and further discussed here. In Chapter VIII it was shown that the SRVAR leads indirectly to an energy-momentum vector that clearly allows for objects with spin $= \frac{1}{2}$. The fundamental equation for conservative motion of a particle with rest mass $= m$ is the energy momentum equation,

$$\mathcal{E}^2 = c^2 p^2 + m^2 c^4 \qquad (10.1)$$

Which implies the existence of a 4-vector,

$$imc^2 \hat{\sigma} = i\mathcal{E}\hat{l} + c\overline{p} = i\gamma mc^2 \hat{l} + \gamma mc^2 \overline{\beta}. \qquad (10.2)$$

Comparing equation (10.2) with equation (9.3), which was obtained from the SRVAL, one sees that if $\gamma\overline{\beta} = (1-d)\overline{a}$ and $\gamma = d$, the correct energy-momentum equation will be obtained for equation (9.3) when it is multiplied by mc^2. A self consistent energy-momentum equation and Lorentz transformation for this case is obtained if and only if $d = \frac{1}{2}$. A possible additional self-consistency requirement is that it should not matter in what order velocities are added. I assume also that the fundamental angular momentum vector is given by, $\overline{l} = \overline{r} \times \overline{p} = \overline{r} \times \gamma mc\overline{a}$. It is also assumed, for simplicity, that the particle is a point mass with point electronic charge ($q = -e$) and all spin motion is in a perfect circle with velocity, $\overline{v} = c\overline{a}$. The angular momentum is then,

$$\overline{l} = \frac{m_e c r_e a}{2} \hat{e} \qquad (10.3)$$ where $\hat{e}$ is the unit vector in the direction of $\overline{l}$. The magnetic moment, $\overline{\mu}_e$, due to this motion of the charge, can be written immediately as

$$\bar{\mu} = -\frac{ecar_e}{2}\hat{e} \quad (10.4)$$

It follows immediately that

$$\bar{\mu} = -\frac{e}{m_e c}\bar{l} \quad (10.5)$$

Therefore, the correct gyromagnetic ratio is obtained. A major hurdle for any electron spin model has been crossed without sacrificing rationality. And what is perhaps more astonishing is that the feat has been accomplished with embarrassingly simple algebra. Also note that a point charge has no dipole moment. The lack of an electric dipole moment is consistent with experimental measurements of the properties of an electron. It remains to determine how a point charge can execute such an orbit and how the singularity in the E-M field is avoided (nature abhors a naked singularity).

A century ago Henri Poincare' showed that mechanical stresses could hold the electron together; i.e. provide stability by compensating the Maxwell stresses and making the total self stress vanish. It was found that the self-energy cannot be divided into an electromagnetic contribution and a mechanical contribution because the separate parts behave differently under Lorentz transformations. Finally, quantum mechanics should enter the picture in the form, at least, of mass, charge, and wavefunction renormalization.

It was shown in Chapter IX, page 60, that a somewhat mysterious 4th component of the force on a particle could or may exist that would exactly compensate the spin centrifugal force in magnitude. This force, F_4 , was given by equation (9.31), namely:

$$iF_4\hat{l} = -\frac{k_0 m_0 c^2}{2}\hat{l} \quad (10.6)$$

This, however, is the magnitude of the force directly at the center of mass of the electric charge (located at radius $r = \frac{1}{k_0}$); and so it's variation in space and time is not explicitly known yet. A guess can only be made at this time; for example, the force could be due a charge q located at $r = 0$, so that,

$$\vec{F}_4 = -\frac{qe}{r^2}\hat{r}$$

$$-\frac{k_0 m_0 c^2}{2} = -qek_0^2 \qquad (10.7)$$

$$q = \frac{m_0 c^2}{2ek_0} = \frac{e}{2\alpha}$$

Where α is the fine structure constant. The charge q would, therefore, necessarily be the same magnitude as the charge of a Dirac magnetic monopole if this guess is true. However, the mass of the Dirac magnetic monopole, which would somehow be "seen" by the electron to be very large, could only be sensed by an observer external to the electron as being equal to, $\frac{m_0}{2}$. The radius $r = \frac{1}{k_0}$ is, perhaps, some sort of "event" horizon cloaking a singularity in the force field. An electron test probe particle perhaps could not go below this horizon radius without exceeding the velocity of light. A probe particle with greater rest mass, m_1, could perhaps penetrate deeper, but only to a new "event" horizon $r = \frac{m_0}{k_0 m_1}$. But we just do not yet know. It may be that below the "event" horizon somehow the quantum mechanical wavefunction takes over and holds the electron to a circular orbit because its phase velocity varies as $v_p = ck_0 r$ so that its refractive index n decreases as, $n = \frac{1}{k_0 r}$. The wavefunction would be guided in a perfectly circular orbit as in an absolute optical instrument. Again we do not know. Perhaps the "event" horizon is the boundary between two different 4-dimensional spaces and the electron has half of its mass in one space and half in the other space. The centripetal force in our space external to the horizon is directed out ward, i.e. in the plus $\bar{r}$ direction and the centripetal force in the internal space is directed in the negative $\bar{r}$ direction so that the forces are balanced. The electric charge $-e$ would be on the external side of the horizon and the electric charge $+e$ on the internal side of the horizon with no communication of the fields across the horizon. At this time we can only make speculations.

The other force equations from Chapter IX were the following:

$$\frac{\partial \mathcal{E}}{\partial t} = c \overline{\beta} \square \overline{F}$$

$$\overline{F} = \frac{\partial \overline{p}}{\partial t} + \nabla \mathcal{E}$$

$$\overline{\beta} \times \overline{F} = 0.$$

Where, $\overline{\beta} = \overline{a}$, is now the orbital velocity of the spinning particle. These three equations are consistent with an external force, $\overline{F}$, acting on the particle. For the development of the spin model, external forces can be set to zero, thus, $\overline{F} = 0$. In this case one gets, $\frac{\partial \mathcal{E}}{\partial t} = 0$ and $\frac{\partial \overline{p}}{\partial t} = -\nabla \mathcal{E}$. If one takes $\nabla \mathcal{E} = \nabla \phi$ where ϕ is the potential energy of the force causing the particle to spin, i.e. the potential corresponding to F_4 and the spin momentum to be, $\overline{p} = \frac{1}{2} m_0 c \overline{a}$, then the equation of motion is

$$\frac{1}{2} m_0 c \frac{\partial \overline{a}}{\partial t} = -\nabla \phi . \qquad (10.8)$$

At this point it becomes necessary to determine what function of $\overline{r}$ and t the potential ϕ is, in order to proceed. Here, a break with tradition is required because one must consider the particle to be some sort of Quantum Mechanical combination of both classical wave and classical particle. Again, more work is necessary.

It has also been shown in Chapter VIII, that rest mass quantization can be achieved in two possible ways. If the energy-momentum vector for an electron is taken to be

$$m_e c \left[i \hat{\sigma} = i d \hat{l} + (1-d) \overline{a} \right] =$$

$$i m_e c \hat{\sigma} = i \gamma m_e c \hat{l} + \gamma m_e c \overline{a} = i \frac{m_e c}{2} \hat{l} + \frac{m_e c}{2} \overline{a}$$

The square of this vector is,

$$-m_e^2 c^4 = -\frac{1}{4} m_e^2 c^4 + \frac{1}{4} m_e^2 c^4 a^2$$

Consequently, as it was indicated in Chapter VIII, if, $a^2 = -3$, then the proper rest mass is obtained. However, this is not too satisfying because $\bar{a}$ must be imaginary and its absolute value greater than 1. Imaginary vectors are allowed in the author's system of mathematics and the SRVAR, taken at face value, gives a velocity sum for

$$\bar{s} = i\hat{i} \oplus i\hat{j} = i\hat{i} + i\hat{j} + i\left(i\hat{i} \times i\hat{j}\right)$$

$$s^2 = -1 - 1 - 1 = -3$$

(The symbol $\oplus$ will frequently be used to designate addition by the SRVAR.) The above is mathematically okay, but as shown in Chapter VIII, it may be better to use the results of equations (9.9) and (9.10) for the values of d. This procedure yields the equations, (9.13), (9.14), (9.17), and (9.18), which give the correct rest mass quantization for $a^2 = 1$. Consider the vector

$$m_0 c\hat{s} = m_0 c\left[\left(\frac{i}{2}\hat{i} + \frac{1}{2}\hat{a}\right) - \left(\frac{1}{2}\hat{i} + \frac{i}{2}\hat{a}\right)\right].$$

The carets above a letter denote that it is a unit vector. This vector could represent the momentum vector for a spin $\frac{1}{2}$ particle with rest mass m_0.

Even more general such vectors can be written; for example, a four component vector can be developed by use of quaternions, thus

$$\bar{s} = d\hat{l} + a\hat{i} + b\hat{j} + c\hat{k}$$ where a, b, c, and d are numbers and $\hat{l}$, $\hat{i}$, $\hat{j}$, and $\hat{k}$ have the following multiplication table:

	$\hat{l}$	$\hat{i}$	$\hat{j}$	$\hat{k}$
$\hat{l}$	$\hat{l}$	$\hat{i}$	$\hat{j}$	$\hat{k}$
$\hat{i}$	$\hat{i}$	$-\hat{l}$	$\hat{k}$	$-\hat{j}$
$\hat{j}$	$\hat{j}$	$-\hat{k}$	$-\hat{l}$	$\hat{i}$
$\hat{k}$	$\hat{k}$	$\hat{j}$	$-\hat{i}$	$-\hat{l}$

The vector $\bar{s}$ then has the property

$$\bar{s} \times \bar{s} = 2d\bar{s} - \left(d^2 + a^2 + b^2 + c^2\right)\hat{i} \quad (10.9)$$

Does this look familiar? This vector is formally the same as the □ matrix. With suitable values for the components, this type vector should be useful for describing the 4-momentum of various particles.

A few things concerning all the energy-momentum vectors need to be cleared up.

One of these is what kind of particle or group of particles could be represented by the vector $\bar{s} = 0\hat{i} + \bar{a}$? Consider the following sums of vectors:

$$\left(id\hat{i} + \bar{a}\right) \oplus \left(id\hat{i} + \bar{a}\right) = 2id\hat{i} + 2(1-d)\bar{a}$$

$$\left(id\hat{i} + \bar{a}\right) \oplus \left(-id\hat{i} - \bar{a}\right) = 2d\bar{a}$$

$$\left(id\hat{i} + \bar{a}\right) \oplus \left(id\hat{i} - \bar{a}\right) = 2id\hat{i}$$

$$\left(id\hat{i} + \bar{a}\right) \oplus \left(-id\hat{i} + \bar{a}\right) = 2\bar{a}$$

(10.10)

Therefore,

$$\bar{s} = 0 + \bar{a} = \frac{1}{2d}\left[\left(id\hat{i} + \bar{a}\right) \oplus \left(-id\hat{i} - \bar{a}\right)\right] ......or... \bar{s} = 0 + \bar{a} = \frac{1}{2}\left[\left(id\hat{i} + \bar{a}\right) \oplus \left(-id\hat{i} + \bar{a}\right)\right].$$

Also one can see that $\bar{s} = \frac{1}{2}\left[\left(id\hat{i} + \bar{a}\right) \oplus \left(id\hat{i} - \bar{a}\right)\right] = id\hat{i}$. The vector, $\bar{s} = 0 + \bar{a}$, would represent a particle with non-zero rest mass, $m_0 = \pm m_e a^2$ (?) and spin, but no 4th component of momentum. However, the third vector represents a particle with non-zero rest mass (m_0) and rest momentum $= m_0 c$, but with zero spin.

Can one gain any insight by considering four dimensional rotations? Consider rotation of a 4-D vector by an angle γ about the z-axis and simultaneous rotation by an angle δ about the time axis. The rotation matrices are given by equations (4.3) and (4.4) respectively multiplied together:

$$R_z = \begin{pmatrix} e^{i\gamma} & 0 & 0 & 0 \\ 0 & \cos\gamma & -\sin\gamma & 0 \\ 0 & \sin\gamma & \cos\gamma & 0 \\ 0 & 0 & 0 & 1 \end{pmatrix}$$

$$R_t = \begin{pmatrix} 1 & 0 & 0 & 0 \\ 0 & e^{i\delta} & 0 & 0 \\ 0 & 0 & e^{i\delta} & 0 \\ 0 & 0 & 0 & e^{i\delta} \end{pmatrix} \quad \textit{which yields the matrix}$$

$$R_z R_t = \begin{pmatrix} e^{i\gamma} & 0 & 0 & 0 \\ 0 & e^{i\delta}\cos\gamma & e^{i\delta}\sin\gamma & 0 \\ 0 & -e^{i\delta}\sin\gamma & e^{i\delta}\cos\gamma & 0 \\ 0 & 0 & 0 & e^{i\delta} \end{pmatrix} \qquad (10.11)$$

However,

$$e^{i\delta}\cos\gamma = \frac{1}{2}\Big[\cos(\delta+\gamma)+\cos(\delta-\gamma)+i\big[\sin(\delta+\gamma)+\sin(\delta-\gamma)\big]\Big]$$

$$and$$

$$e^{i\delta}\sin\gamma = \frac{1}{2}\Big[\sin(\delta+\gamma)-\sin(\delta-\gamma)-i\big[\cos(\delta+\gamma)-\cos(\delta-\gamma)\big]\Big]$$

Now, if $\delta = \gamma = \theta$, one can write

$$R_z R_t = \begin{pmatrix} e^{i\theta} & 0 & 0 & 0 \\ 0 & \frac{1}{2}\left(1+e^{2i\theta}\right) & \frac{i}{2}\left(1-e^{2i\theta}\right) & 0 \\ 0 & \frac{i}{2}\left(1-e^{2i\theta}\right) & \frac{1}{2}\left(1+e^{2i\theta}\right) & 0 \\ 0 & 0 & 0 & e^{i\theta} \end{pmatrix} \qquad (10.12)$$

In this case, the x and y components of the vector would be turning at twice the angle of the z and time components. This fits our spin model if the frequency of rotation is given by $\omega_0 = \dfrac{3m_0c^2}{4\hbar}$.

This is a very high frequency, of the order of 10^{21} Hz, and therefore unlikely to be detected directly. Indeed it is more likely that such an oscillation would be averaged to zero in any experimental measurement. It follows that the time average of matrix is what would be observed, thus

$$\langle R_z R_t \rangle = \begin{pmatrix} 0 & 0 & 0 & 0 \\ 0 & \dfrac{1}{2} & \dfrac{i}{2} & 0 \\ 0 & -\dfrac{i}{2} & \dfrac{1}{2} & 0 \\ 0 & 0 & 0 & 0 \end{pmatrix}. \qquad (10.13)$$

The possible other time-averaged rotation matrices similar to equation are

$$\langle R_x R_t \rangle = \begin{pmatrix} 0 & 0 & 0 & 0 \\ 0 & \dfrac{1}{2} & 0 & -\dfrac{i}{2} \\ 0 & 0 & 0 & 0 \\ 0 & \dfrac{i}{2} & 0 & \dfrac{1}{2} \end{pmatrix} \qquad (10.14)$$

$$\langle R_y R_t \rangle = \begin{pmatrix} 0 & 0 & 0 & 0 \\ 0 & \dfrac{1}{2} & 0 & -\dfrac{i}{2} \\ 0 & 0 & 0 & 0 \\ 0 & \dfrac{i}{2} & 0 & \dfrac{1}{2} \end{pmatrix} \qquad (10.15)$$

Clearly the time averaged matrices represent pseudo 4-D or 3-D spaces and are really more two dimensional than anything else. This seems to be telling us that each component of the spin is two dimensional on average and that time is a sort of square root space. You will notice that all of these matrices are singular matrices.

A wavefunction that would be applicable to the spin model is now constructed following the method discovered in Chapter VII. One simply uses the appropriate Lorentz-transformed position 4-vector:

$$\psi = \psi_0 \exp\left[-i\gamma k \left(\overline{a} \square \overline{r} - ct \right) - \gamma \overline{k} \square \left(\overline{r} - \overline{a}ct \right) + i\gamma \left(\overline{k} \times \overline{a} \right) \square \overline{r} \right] (10.16)$$

where $\overline{a} \perp \overline{r}$, $\overline{k} = \dfrac{\overline{k}_0 m}{m_0}$, $\hbar = \dfrac{m_0 c}{k_0}$, and, of course, $\gamma = \dfrac{1}{2}$ and

therefore

$$\psi = \psi_0 \exp\left[i\frac{1}{2}kct - \frac{1}{2}\overline{k}\square(\overline{r} - \overline{a}ct) + i\frac{1}{2}\left(\overline{k} \times \overline{a}\right)\square\overline{r} \right] \quad (10.17)$$

Applying the momentum operator $-i\hbar\nabla$ and the energy operator $i\hbar\dfrac{\partial}{\partial t}$ to, ψ, one obtains the following results:

$$-i\hbar\nabla\psi = \overline{p}\psi$$

$$i\hbar\frac{\partial\psi}{\partial t} = \mathcal{E}\psi$$

$$(10.18)$$

$$\overline{p} = \frac{imc}{2}\hat{k} + \frac{mca}{2}\left(\hat{k} \times \hat{a}\right)$$

$$\mathcal{E} = +\frac{imc^2 a}{2} - \frac{mc^2}{2}$$

But $\hat{k}\square\hat{a}$ and therefore, $\hat{k} \times \hat{a} = 0$, consequently

$$\overline{p} = \frac{imc}{2}\hat{k}$$

$$\mathcal{E} = +\frac{imc^2 a}{2} - \frac{mc^2}{2}$$

$$(10.19)$$

The correct energy value is obtained if, $a = i\sqrt{3}$, as I have found before in this simple case.

Chapter XI. Further Development of Ideas – Grand Equations of Motion

Equation Section (Next)

In Chapter V it was shown that the D operator acting on the 4-momentum vector (see equation (5.26)) leads to two of Hamilton's equations of motion (pp 23 - 24). In Chapter VII it was shown that, if we require the vorticity of the momentum to be zero and the entity undergoing motion to be something which has the attributes of both a wave and a particle, then a rudimentary quantum mechanics obtains. Let us go one step farther along this path. First, because this will involve 4-vectors, let us recall the fundamental 4-vectors that are at our disposal. These are the energy-momentum vector, the position 4-vector, the velocity 4-vector, the acceleration vector, the force vector, and the angular momentum vector. Already we are faced with two of the ambiguities of relativity. It is not clear what the acceleration, force, and angular momentum vectors should be or that they are even 4-vectors. Indications from our results in prior chapters and from experiment are that there is an intrinsic spin or rest angular momentum for elementary particles (and thus angular momentum should be a 4-vector); but acceleration may not have a non-zero rest component. Let us assume that the basic relativistic force vector is given by $iF_4\hat{l} + \overline{F}$. To get a reasonable 4-force, a Lorentz transform is applied to this vector, thus

$$\gamma \begin{pmatrix} 1 & -i\overline{\beta} \\ i\overline{\beta} & 1 \end{pmatrix} \begin{pmatrix} iF_4\hat{l} \\ \overline{F} \end{pmatrix} = \begin{pmatrix} i\gamma\left(F_4 - \overline{\beta}\square\overline{F}\right)\hat{l} \\ \gamma\left(\overline{F} - F_4\overline{\beta}\right) - i\gamma\overline{\beta}\times\overline{F} \end{pmatrix} \qquad (11.1)$$

Generalizing this result by multiplying by the function, $\psi(\overline{r},ct)$, taking note of the results obtained in Chapter IX, pages 57-58, equation (9.25), whereby the differential operator, D, is made to look like a Lorentz transformation, and generalizing the energy-momentum 4-vector by multiplying it with a function, $\psi(\overline{r},ct)$, we operate on the energy-momentum 4-vector with γ times the differential $\Box$ matrix and set the result equal to ψ times the Lorentz transformed force vector to get

$$\gamma \begin{pmatrix} \dfrac{1}{c}\dfrac{\partial}{\partial t} & -i\nabla \\[2mm] i\nabla & \dfrac{1}{c}\dfrac{\partial}{\partial t} \end{pmatrix} \begin{pmatrix} -i\mathcal{E}\psi\hat{i} \\[2mm] c\bar{p}\psi \end{pmatrix} = -i\gamma \left[\left(\dfrac{1}{c}\dfrac{\partial \mathcal{E}}{\partial t} + c\nabla\Box\bar{p} \right)\psi + \left(\dfrac{\mathcal{E}}{c}\dfrac{\partial\psi}{\partial t} + c\bar{p}\Box\nabla\psi \right) \right]\hat{i}$$

$$+\gamma \left[\left(\nabla\mathcal{E} + \dfrac{\partial\bar{p}}{\partial t} \right)\psi + \mathcal{E}\nabla\psi + \bar{p}\dfrac{\partial\psi}{\partial t} \right] + \gamma\left[-ic\left(\nabla\times\bar{p}\right)\psi - ic\bar{p}\times\nabla\psi \right] =$$

$$i\gamma\left(F_4 - \bar{\beta}\Box\bar{F} \right)\psi\hat{i} + \gamma\left(\bar{F} - F_4\bar{\beta} \right)\psi - i\gamma\left(\bar{\beta}\times\bar{F} \right)\psi$$

$$(11.2).$$

This is the grand equation of motion provided $\mathcal{E}$ is the total energy including any energy of rotation, $\bar{p}$ is the total momentum including any angular momentum, and $\bar{F}$ is the total force including any torques acting on the particle. The grand equation can be written as two equations:

$$-i\gamma\left[\left(\dfrac{1}{c}\dfrac{\partial\mathcal{E}}{\partial t} + c\nabla\Box\bar{p} \right)\psi + \left(\dfrac{\mathcal{E}}{c}\dfrac{\partial\psi}{\partial t} + c\bar{p}\Box\nabla\psi \right) \right] = i\gamma\left(F_4 - \bar{\beta}\Box\bar{F} \right)\psi \quad (11.3)$$

$$\left(\nabla\mathcal{E} + \dfrac{\partial\bar{p}}{\partial t} \right)\psi + \mathcal{E}\nabla\psi + \bar{p}\dfrac{\partial\psi}{\partial t} - ic\left(\nabla\times\bar{p}\right)\psi - ic\bar{p}\times\nabla\psi =$$
$$+\left(\bar{F} - F_4\bar{\beta} \right)\psi - i\left(\bar{\beta}\times\bar{F} \right)\psi \quad (11.4)$$

Applying the quantum mechanical operators,
$$-i\hbar\nabla\psi = \bar{p}\psi$$

$$i\hbar\dfrac{\partial}{\partial t}\psi = \mathcal{E}\psi$$
to equations (11.3) through(11.4), one gets from equation (11.3) the result:

$$-\dfrac{1}{c}\dfrac{\partial\mathcal{E}}{\partial t} - c\nabla\Box\bar{p} + i\dfrac{\mathcal{E}^2}{\hbar c} - i\dfrac{c^2 p^2}{\hbar c} = F_4 - \bar{\beta}\Box\bar{F} \quad (11.5) \quad \text{which}$$

splits into the equations:

$$\dfrac{1}{c}\dfrac{\partial\mathcal{E}}{\partial t} = \bar{\beta}\Box\bar{F} \quad (11.6)$$

$$F_4 = i\dfrac{\mathcal{E}^2}{\hbar c} - i\dfrac{c^2 p^2}{\hbar c} = i\dfrac{m_0^2 c^4 k_0}{m_0 c^2} = im_0 c^2 k_0 \quad (11.7)$$

And, of course, $\nabla \overline{\Box p} = 0$. Equation (11.4) becomes:

$$\nabla \mathcal{E} + \frac{\partial \overline{p}}{\partial t} + i\frac{\mathcal{E}p}{\hbar} - i\frac{\mathcal{E}p}{\hbar} = +\overline{F} - F_4 \overline{\beta} - i\overline{\beta} \times \overline{F} \quad (11.8)$$

Or

$$\overline{F} = \nabla \mathcal{E} + \frac{\partial \overline{p}}{\partial t} + im_0 c^2 k_0 \overline{\beta} + i\overline{\beta} \times \overline{F} \quad (11.9)$$

Equation (11.9) is problematical and could be incorrect, because not all angular momenta have been included in the momentum and not all torques in the force. Note that if the particle's rest mass is zero, $F_4 = 0$, and the nature of the problem changes somewhat. It seems that it will be necessary to determine what the angular momenta for a massive particle should be and whether part of the angular momentum is missing from equation (11.9). Equation (11.6) is a statement that the rate of change of energy of the particle is equal to the work done by the force $\overline{F}$ acting on the particle and is therefore correct. Equation (11.7) appears to be all right, except that the nature of F_4 is at the moment a bit mysterious.

If one uses an argument for the wavefunction ψ based on the Lorentz transformed position vector, the quantum mechanical operator equations become:

$$\nabla \psi = \left[-i\gamma k_4 \overline{\beta} - \gamma \overline{k} + i\gamma \left(\overline{k} \times \overline{\beta} \right) \right] \psi$$

$$\frac{\partial \psi}{\partial t} = \left[i\gamma k_4 c + \gamma \left(\overline{k} \Box \overline{\beta} \right) c \right] \psi$$

where

$$\psi = \psi_0 \exp \left[-i\gamma k_4 \left(\overline{\beta} \Box \overline{r} - ct \right) - \gamma \overline{k} \Box \left(\overline{r} - \overline{\beta} ct \right) + i\gamma \left(\overline{k} \times \overline{\beta} \right) \Box \overline{r} \right]$$

This wavefunction yields essentially the same results as before, namely the equations:

$$\overline{\beta} \Box \overline{F} = \frac{1}{c} \frac{\partial \mathcal{E}}{\partial t} + c\nabla \Box \overline{p} \quad (11.10)$$

$$F_4 = im_0 c^2 k_4 \quad (11.11)$$

$$\overline{F} - i\overline{\beta} \times \overline{F} = \nabla \mathcal{E} + \frac{\partial \overline{p}}{\partial t} + ic\nabla \times \overline{p} \quad (11.12)$$

$$F_4 \overline{\beta} = m_0 c^2 \overline{k} \quad (11.13)$$

This is better, but not very satisfying because $\bar{k}$ must equal $ik_4\bar{\beta}$ and, therefore the $i\gamma\left(\bar{k}\times\bar{\beta}\right)\Box\bar{r}$ term in the argument of ψ must then be zero. Evidently, one must turn to a spin model to solve this problem.

Let us look at equations (11.3) and (11.4) in a different way. Instead of expressing them in terms of the expectation or eigenvalues of the quantum mechanical operators, let us write them in terms of the operators, thus,

$$-i\gamma\left[\left(\frac{1}{c}\frac{\partial\mathcal{E}}{\partial t}+c\nabla\Box\bar{p}\right)\psi+\left(\frac{\mathcal{E}}{c}\frac{\partial\psi}{\partial t}+c\bar{p}\Box\nabla\psi\right)\right]=$$

$$\gamma\hbar c\left(\frac{1}{c^2}\frac{\partial^2\psi}{\partial t^2}-\nabla^2\psi\right)=\left(function?\right)\psi$$

(11.14)

Equation (11.4) becomes

$$\left(\nabla\mathcal{E}+\frac{\partial\bar{p}}{\partial t}\right)\psi+\mathcal{E}\nabla\psi+\bar{p}\frac{\partial\psi}{\partial t}-ic\left(\nabla\times\bar{p}\right)\psi-ic\bar{p}\times\nabla\psi=$$

$$i\hbar\left(\nabla\frac{\partial\psi}{\partial t}-\frac{\partial}{\partial t}\nabla\psi\right)-i\hbar c\nabla\times\nabla\psi=0$$

(11.15)

This suggests that equation (11.2) should be set equal to the vector $iF_4\hat{l}+\bar{F}$ with $\bar{F}=0$. This gives

$$\nabla^2\psi-\frac{1}{c^2}\frac{\partial^2\psi}{\partial t^2}=-i\frac{F_4}{\hbar c}\psi=-k_4^2\psi \text{ (11.16), that is, essentially the}$$

Klein-Gordon equation. This not perfect, but it is perhaps better than before.

Now we can proceed as above, but instead of using differentials with respect to the coordinates in the differential operator, D, we use differentials with respect to the conjugate momenta, i.e. the energy divided by $c = p_4$ and the momentum, $\bar{p}$. Such an approach is with an eye toward introducing suitably generalized coordinates and momenta and then a configurational space. As for space, it appears to be necessary in the present scheme to introduce an eight dimensional space to accommodate both the generalized coordinates and the generalized momenta (time, plus three space coordinates, plus four momenta). For this book, I must stop here or it will never be finished. Nevertheless, I will continue to work on these matters and reserve them for a future edition.

Chapter XII.   Recapitulation

Equation Section (Next)

The material covered in the first seven chapters builds a case for the validity and usefulness of a particular type of mathematical operator in developing fundamental theoretical physics: namely, the nearly periodic ▢ matrix operator, e.g.

$$\Box_1 = \begin{pmatrix} d & a & b & c \\ -a & d & -c & b \\ -b & c & d & -a \\ -c & -b & a & d \end{pmatrix} \qquad (3.1)$$

The $\Box_1$ matrix is constructed from the 4-vector, $d\hat{l} + a\hat{i} + b\hat{j} + c\hat{k}$ in a fairly transparent manner by substituting the 4-vector components for the $\Box_1$ matrix elements as shown. In the same manner components the velocity 4-vector, $ic\hat{\sigma} = i\gamma c\hat{l} + \gamma c\bar{\beta}$, are used to construct a Lorentz transformation matrix, dividing the whole matrix by ic, the magnitude of the 4-velocity, to get

$$\mathcal{L} = \gamma \begin{pmatrix} 1 & -i\beta_x & -i\beta_y & -i\beta_z \\ i\beta_x & 1 & -i\beta_z & i\beta_y \\ i\beta_y & i\beta_z & 1 & -i\beta_x \\ i\beta_z & -i\beta_y & i\beta_x & 1 \end{pmatrix} \qquad (5.2).$$

This Lorentz transform is generalized, yields correct results in every case, and is very much simpler in form and in use than the conventional Lorentz transform matrix.

Two successive Lorentz transforms, can be shown to give the following matrix

$$\gamma_\alpha \begin{pmatrix} 1 & -i\alpha_x & -i\alpha_y & -i\alpha_z \\ i\alpha_x & 1 & i\alpha_z & -i\alpha_y \\ i\alpha_y & -i\alpha_z & 1 & i\alpha_x \\ i\alpha_z & i\alpha_y & -i\alpha_x & 1 \end{pmatrix} \gamma_\beta \begin{pmatrix} 1 & -i\beta_x & -i\beta_y & -i\beta_z \\ i\beta_x & 1 & i\beta_z & -i\beta_y \\ i\beta_y & -i\beta_z & 1 & i\beta_x \\ i\beta_z & i\beta_y & -i\beta_x & 1 \end{pmatrix} =$$

$$\gamma_\alpha\gamma_\beta \begin{pmatrix} 1+\overline{\alpha}\square\overline{\beta} & -i(\alpha_x+\beta_x)-(\overline{\alpha}\times\overline{\beta})_x & -i(\alpha_y+\beta_y)-(\overline{\alpha}\times\overline{\beta})_y & -i(\alpha_z+\beta_z)-(\overline{\alpha}\times\overline{\beta})_z \\ i(\alpha_x+\beta_x)+(\overline{\alpha}\times\overline{\beta})_x & 1+\overline{\alpha}\square\overline{\beta} & i(\alpha_z+\beta_z)+(\overline{\alpha}\times\overline{\beta})_z & -i(\alpha_y+\beta_y)-(\overline{\alpha}\times\overline{\beta})_y \\ i(\alpha_y+\beta_y)+(\overline{\alpha}\times\overline{\beta})_y & -i(\alpha_z+\beta_z)-(\overline{\alpha}\times\overline{\beta})_z & 1+\overline{\alpha}\square\overline{\beta} & i(\alpha_x+\beta_x)+(\overline{\alpha}\times\overline{\beta})_x \\ i(\alpha_z+\beta_z)+(\overline{\alpha}\times\overline{\beta})_z & i(\alpha_y+\beta_y)+(\overline{\alpha}\times\overline{\beta})_y & -i(\alpha_x+\beta_x)-(\overline{\alpha}\times\overline{\beta})_x & 1+\overline{\alpha}\square\overline{\beta} \end{pmatrix}$$

If we factor the term $1+\overline{\alpha}\square\overline{\beta}$ out of the matrix, the matrix will then have diagonal elements equal to 1 and a gamma equal to $\gamma_s = \gamma_\alpha\gamma_\beta(1+\overline{\alpha}\square\overline{\beta})$. Consequently, it would resemble a Lorentz transform matrix for the sum of velocities given by

$$\overline{s} = \frac{\overline{\alpha} + \overline{\beta} - i(\overline{\alpha}\times\overline{\beta})}{1+\overline{\alpha}\square\overline{\beta}}$$

One could have achieved the same result by performing two successive transformations on the time-position 4-vector. The sign of the cross product term does not seem to matter if the sum is always done in a consistent manner. In my book, I have chosen it to be positive, so that

$$\overline{s} = \frac{\overline{\alpha} + \overline{\beta} + i(\overline{\alpha}\times\overline{\beta})}{1+\overline{\alpha}\square\overline{\beta}},$$

In most calculations, it is the magnitude of $\overline{s}$,

$s = \sqrt{\alpha^2 + \beta^2 - \alpha^2\beta^2}$, which is important, so the sign of the cross term does not matter. This question must, of course, eventually be completely resolved.

We see then that this Lorentz transform leads, in turn, to a generalized velocity addition rule for special relativity (SRVAR),

$$\overline{s} = \frac{\overline{\alpha} + \overline{\beta} + i(\overline{\alpha}\times\overline{\beta})}{1+\overline{\alpha}\square\overline{\beta}}$$ (5.16) which not only gives correct

results, but is simple and easily applied to any arbitrary velocity 4-vector. The SRVAR leads in its turn, although not so directly, to a relativistic energy momentum vector, for an elementary particle with non-zero rest mass, which requires a spin of $+\frac{1}{2}$; i.e. the gamma for the Lorentz transformation must be $\frac{1}{2}$ regardless of what velocity 3-vector the particle may have or what rest mass the particle may have. This leads directly to a spin model for an electron that yields the correct gyromagnetic ratio without any overtly irrational features. This result leads, in its turn, to correct quantization of the rest mass of an elementary particle.

And yet further, the $\square$ matrix leads directly to a differential form of this type matrix, constructed from the differential operator 4-vector, $\mathcal{D} = i\frac{1}{c}\frac{\partial}{\partial t}\hat{l} + \nabla$, thus

$$D_1^* = \begin{pmatrix} \dfrac{i}{c}\dfrac{\partial}{\partial t} & \dfrac{\partial}{\partial x} & \dfrac{\partial}{\partial y} & \dfrac{\partial}{\partial z} \\[2ex] -\dfrac{\partial}{\partial x} & \dfrac{i}{c}\dfrac{\partial}{\partial t} & -\dfrac{\partial}{\partial z} & \dfrac{\partial}{\partial y} \\[2ex] -\dfrac{\partial}{\partial y} & \dfrac{\partial}{\partial z} & \dfrac{i}{c}\dfrac{\partial}{\partial t} & -\dfrac{\partial}{\partial x} \\[2ex] -\dfrac{\partial}{\partial z} & -\dfrac{\partial}{\partial y} & \dfrac{\partial}{\partial x} & \dfrac{i}{c}\dfrac{\partial}{\partial t} \end{pmatrix} \qquad (3.12)$$

When D_1^* is multiplied into the electromagnetic vector potential, $i\phi\hat{l} + \overline{A}$, the electromagnetic fields, $i\overline{E} + \overline{B}$, are obtained. When the complex conjugate, D_1 , is multiplied into the E-M field, $i\overline{E} + \overline{B}$, and the result set equal to the charge-current density 4-vector, the complete set of Maxwell's equations are obtained in one step. Operation again with D_1^* on the Maxwell equations or on the charge-current density 4-vector yields the charge-current density conservation law in 4 dimensional form. Multiplication of D_1 on the charge-current density conservation law yields wave equations for the charge density and for the current density. One will note that this sequence of operations decreases the physical units of the quantities by a factor of $\dfrac{1}{centimeter}$ at each step. This sequence in dimensions is illustrated by:

102

Energy per unit charge → force per unit charge → charge per cubic cm → energy density per unit charge → force density per unit charge.

Demonstrated elsewhere by the author, but not yet in this book, an electromagnetic stress-energy tensor can be constructed from the electromagnetic field 4-vector and the ☐ matrix. This matrix, operated on by the D_1 matrix or its complex conjugate and set equal to the result of multiplying D_1 on the mechanical energy–momentum vector, yields the Lorentz force, the Poynting vector, and virtually all the rest of the equations of fundamental importance in electromagnetic theory.

Finally, to complete the tour de force of electromagnetism, it is shown with the aid of the D_1^* matrix (slightly modified to take into account retardation effects) multiplied by the Lienard-Wiechert potentials (in four vector form) yields the E-M field in a 4-vector form that contains the correct three dimensional electromagnetic fields of a charge under arbitrary motion - along with the surprising result that the fourth or time component of the electromagnetic field is automatically zero. That is, there is no motion of a charge that will give rise to a fourth component of force. This result, coupled with the earlier result that the Lorentz condition and gauge invariance of the potentials makes the fourth component of the fields zero, means that electromagnetism does not have any force in the time direction. If this is true of all forces with gauge invariant force fields, then there may not be a fourth component of the gravitational field, the only other long range force. If there is no direct way to sense anything in the time direction, then we creatures here on earth have no way of knowing which way is the time direction.

The vector obtained by operation of the D_1^* matrix on the energy-momentum 4-vector yields a force equation of a particle, which is also rather like a continuity equation for energy and momentum. If one eschews the pure particle picture of the entity we call an elementary particle and ascribe some wave nature to it, then the curl (vorticity) of the momentum can be made equal to zero by requiring the momentum to be the gradient of some function just as well as requiring the momentum to exist only at the center of mass. This line of thinking plus the force 4-vector yields the conventional

Schroedinger type operators of quantum mechanics for energy and momentum; and as a bonus the Klein Gordon equation. (To get Planck's constant from first principles, the spin model must be further developed.)

It is found that the dot product of the Lorentz transformed position 4-vector with a suitable propagation vector is an excellent argument for the quantum mechanical wavefunction of a free particle. The particle can be localized or not localized to any reasonable degree, has both group and phase velocities explicitly represented, has a purely oscillatory part, and has what appears to be a spin wavefunction part.

Use of the special relativity velocity addition rule leads one to momentum 4-vectors, which represent the momenta of particles with spin $\frac{1}{2}$ and properly quantized non-zero rest masses. Non-zero rest mass requires spin $\frac{1}{2}$, but not necessarily vice versa. These vectors may be useful in describing the interactions of elementary particles. Much work with this vector formulation remains to be done.

Noteworthy results of lesser importance are listed here:

1. It was found that Dirac matrices are also nearly periodic matrices (Pauli matrices are periodic).

2. An interesting property of $\square$ matrices is that operation on an arbitrary vector of n dimensions with an n dimensional $\square$ matrix yields a vector, which, when squared, has cross terms that sum to zero. This operation can be repeated indefinitely with the same result. That particular property is what makes an otherwise trivial matrix very important, because it fits in so well with the properties of relativistic 4-vectors. The $\square$ matrix is a tensor, is nearly periodic, and has a determinant equal to the sum of the squares of all the different matrix elements raised to the $\frac{n}{2}$ power, where n is the number of dimensions of the matrix. If all off-diagonal elements are pure imaginary and all diagonal elements are pure real numbers, then $\square$ is Hermitian. If all diagonal elements are pure imaginary and all off-diagonal elements are pure real numbers, then $\square$ is Anti-Hermitian.

104

3. The Galilean transform is shown to follow directly from the approximate generalized Lorentz transform obtained when all velocities are much smaller than the velocity of light and thus retardation effects are negligible. The infinitesimal Lorentz transformation is obtained from the generalized Lorentz transformation merely by setting, $\gamma = 1$.

4. Derivations of two of Hamilton's canonical equations are obtained by operation of a D matrix, constructed from derivatives, with respect to the generalized coordinates or to the canonical momenta, on the relativistic energy-momentum 4-vector.

Novel rotation matrices are described in Chapter IV for rotations about the space axes and the time axis. These matrices are derived from the $\square$ matrix and may turn out to have important application for helicity of, for example, neutrinos. The neutrino apparently has non-zero rest mass (due to spontaneous flavor changing). A consequence of non-zero rest mass is that neutrinos very likely must travel at less than the velocity of light. Therefore, there would exist a Lorentz frame, in which any neutrino must be seen to travel backward and thus have right handed helicity in contradiction to observations.

A result, which is not of lesser importance, but which I have not completely substantiated to a satisfactory degree, stems from the SRVAR and bears on the effects of gravitational forces on light and material objects. Three examples are explored in the next Chapter, namely: central force motion of a planet in a closed orbit, central force motion of light in an open orbit, and motion of light in a straight, short radial line in a central force field.

Chapter XIII Gravitation (An unexpected result concerning space curvature) and Conclusion

Equation Section (Next)

Consider the motion of the planet Mercury in its orbit about the sun. This motion is hardly relativistic because the total orbital reduced velocity squared is $\frac{v^2}{c^2} \leq 4 \times 10^{-8}$ meters per second and therefore $\gamma \cong 1$ to very good approximation. Nevertheless, a treatment of this problem with methods of special relativity is deemed necessary, because all the relativistic correction factors, including those that might contribute to perihelion drift, are of the same order of magnitude. Angular momentum is, of course, conserved and the first integral of the motion is given by the relativistic energy-momentum equation:

$$im_0 c^2 \hat{\sigma} = i\mathcal{E}\hat{I} + c\overline{p} \qquad (13.1)$$

Where $\overline{p} = \gamma m_0 c \overline{\beta}$ and $\mathcal{E}$ is the "total" energy. The energy $\mathcal{E}$ is ambiguous, as I have pointed out before, i.e., $\mathcal{E} = \gamma m_0 c^2$, which does not include the potential energy of the gravitational field, but does include the rest energy, $m_0 c^2$. How can one handle this? In this case of closed orbital motion, the most straightforward way makes use of the relation between the orbital eccentricity ε and the energy, E, namely:

$$\varepsilon^2 = 1 + \frac{2EJ^2}{mG^2 M^2} \qquad (13.2)$$

Or

$$E = -\frac{\left(1-\varepsilon^2\right)G^2 M^2 m}{2J^2} \qquad (13.3)$$

Substituting, $J^2 = GMa\left(1-\varepsilon^2\right)$, where a is the semimajor axis of the orbit, one gets,

$$E = -\frac{GMm}{2a} \quad (13.4)$$

Now, however,

$$E = K.E. + P.E. . \quad (13.5)$$

That is, E is equal to the kinetic energy plus the potential energy, but not including the rest energy, which cannot have a direct effect on the motion of the planet. Further, I assert that the kinetic energy is $K.E. = (\gamma - 1)mc^2$, mainly because it is the only energy in the system, which depends entirely on velocity (i.e., in the sense, if velocity is zero this energy is also zero). These results are now inserted into equation, (13.1), thus,

$$imc^2\hat{\sigma} = i\left(\frac{GMm}{r} - \frac{GMm}{2a} + mc^2\right)\hat{i} + \gamma mc^2\bar{\beta} \quad (13.6)$$

Dividing through by mc^2 and changing all velocities to derivatives with respect to proper time, τ, a stratagem employed also by all researchers in Special and General Relativity, one gets

$$i\hat{\sigma} = i\left(\frac{GM}{c^2 r} - \frac{GM}{2c^2 a} + 1\right)\hat{i} + \bar{\beta}' \quad (13.7)$$

Hereafter, the prime on β will be dropped and derivatives with respect to proper time will be understood. Squaring equation (13.7), and then collecting and rearranging terms, one gets,

$$-1 = -\frac{G^2 M^2}{c^4 r^2} - \frac{G^2 M^2}{4c^4 a^2} - 1 + \frac{G^2 M^2}{c^4 ar} - \frac{2GM}{c^2 r} + \frac{GM}{c^2 a} + \beta^2$$

$$\beta^2 = \frac{2GM}{c^2 r} - \frac{G^2 M^2}{c^4 ar} + \frac{G^2 M^2}{c^4 r^2} - \frac{GM}{c^2 a} + \frac{G^2 M^2}{4c^4 a^2}$$

$$(13.8)$$

$$\frac{1}{c^2}\left(\frac{\partial r}{\partial \tau}\right)^2 + \frac{r^2}{c^2}\left(\frac{\partial \theta}{\partial \tau}\right)^2 = \frac{2GM}{c^2 r}\left(1 - \frac{GM}{2c^2 a}\right) + \frac{G^2 M^2}{c^4 r^2} - \frac{GM}{c^2 a} + \frac{G^2 M^2}{4c^4 a^2}$$

$$\left(\frac{\partial r}{\partial \tau}\right)^2 + \frac{J^2}{r^2} = \frac{2GM}{r}\left(1 - \frac{GM}{2c^2 a}\right) + \frac{G^2 M^2}{c^2 r^2} - \frac{GM}{c^2 a} + \frac{G^2 M^2}{4c^4 a^2}$$

$$\left(\frac{\partial r}{\partial \tau}\right)^2 + \frac{J^2}{r^2}\left(1 - \frac{G^2 M^2}{c^2 J^2}\right) = \frac{2GM}{r}\left(1 - \frac{GM}{2c^2 a}\right) - \frac{GM}{c^2 a}\left(1 - \frac{GM}{4c^2 a}\right)$$

$$(13.9)$$

Where the last term on the right immediately above is a constant w.r.t., r, θ, and t. Now the quantities $\dfrac{G^2 M^2}{c^2 J^2} \cong 2.56 \times 10^{-8}$, $\dfrac{GM}{2c^2 a} \cong 1.23 \times 10^{-8}$, $\dfrac{GM}{4c^2 a} \cong 6.12 \times 10^{-9}$ are quite small so that they can be neglected w.r.t. 1, and also $\gamma \cong 1 + \dfrac{v^2}{2c^2} \cong 1 + 2 \times 10^{-8}$ is very nearly equal to 1. Thus, one obtains

$$\left(\frac{\partial r}{\partial t}\right)^2 + \left(r\frac{\partial \theta}{\partial t}\right)^2 = \frac{2GM}{r} - \frac{GM}{a} \qquad (13.10)$$

Where $v_\theta^2 = \left(r\dfrac{\partial \theta}{\partial t}\right)^2 = \dfrac{J^2}{r^2}$. This equation is the well-known Newtonian result, leading to an elliptical orbit.

Returning to the relativistic case, equation, (13.9), I submit that the SRVAR applies in this case so that this equation should have a term given by a cross product of the velocities, that is,

$$\left(\frac{i}{c}\overline{v}_r \times \overline{v}_\theta\right)^2 \qquad (13.11)$$

The Newtonian result equation (13.10) can be used to obtain a very good approximation to the value of the velocities, both taken at the same point , $\overline{r}$, in equation (13.11), thus

$$\left(\frac{i}{c}\overline{v}_r \times \overline{v}_\theta\right)^2 = \left(\frac{i}{c}\left[\left(\frac{2GM}{r} - \frac{GM}{a}\right)^{\frac{1}{2}} \hat{i} - \left(\frac{J}{r}\right)\hat{\theta}\right] \times \left(\frac{J}{r}\right)\hat{\theta}\right)^2$$

$$\qquad (13.12)$$

$$= \left(\frac{i}{c}\left(\frac{2GM}{r} - \frac{GM}{a}\right)^{\frac{1}{2}} \hat{i} \times \left(\frac{J}{r}\right)\hat{\theta}\right)^2$$

Finally we get

$$\left(\frac{i}{c}\overline{v}_r \times \overline{v}_\theta\right)^2 \cong -\frac{2GM}{r}\frac{J^2}{c^2 r^2} + \frac{GM}{a}\frac{J^2}{c^2 r^2} = -\frac{2GMJ^2}{c^2 r^3} + \frac{GM}{2a}\frac{J^2}{c^2 r^2}$$

$$(13.13)$$

Adding this term to equation (13.9), one gets,

$$\left(\frac{\partial r}{\partial \tau}\right)^2 + \frac{J^2}{r^2}\left(1 - \frac{G^2 M^2}{c^2 J^2}\right) - \frac{2GMJ^2}{c^2 r^3} + \frac{GM}{a}\frac{J^2}{c^2 r^2}$$

$$= \frac{2GM}{r}\left(1 - \frac{GM}{2c^2 a}\right) - \frac{GM}{c^2 a}\left(1 - \frac{GM}{4c^2 a}\right)$$

$$\left(\frac{\partial r}{\partial \tau}\right)^2 = \frac{2GM}{r}\left(1 - \frac{GM}{2c^2 a}\right) - \frac{J^2}{r^2}\left(1 - \frac{G^2 M^2}{c^2 J^2} + \frac{GM}{c^2 a}\right) + \frac{2GMJ^2}{c^2 r^3} - \frac{GM}{c^2 a}\left(1 - \frac{GM}{4c^2 a}\right) \quad (13.14)$$

$$= \frac{2GM}{r}\left(1 - \frac{GM}{2c^2 a}\right) - \frac{J^2}{r^2}\left(1 - \frac{GM}{c^2 a\left(1 - \varepsilon^2\right)} + \frac{GM}{c^2 a}\right) + \frac{2GMJ^2}{c^2 r^3} - \frac{GM}{c^2 a}\left(1 - \frac{GM}{4c^2 a}\right)$$

$$= \frac{2GM}{r}\left(1 - \frac{GM}{2c^2 a}\right) - \frac{J^2}{r^2}\left(1 - \frac{GM}{c^2 a\left(1 - \varepsilon^2\right)} + \frac{GM}{c^2 a}\right) + \frac{2GMJ^2}{c^2 r^3} - \frac{GM}{c^2 a}\left(1 - \frac{GM}{4c^2 a}\right)$$

$$= \frac{2GM}{r}\left(1 - \frac{GM}{2c^2 a}\right) - \frac{J^2}{r^2}\left(1 - \frac{GM\varepsilon^2}{c^2 a\left(1 - \varepsilon^2\right)}\right) + \frac{2GMJ^2}{c^2 r^3} - \frac{GM}{c^2 a}\left(1 - \frac{GM}{4c^2 a}\right)$$

Renormalizing   the   angular   momentum   to

$$J_n^2 = J^2\left(1 - \frac{GM\varepsilon^2}{c^2 a\left(1 - \varepsilon^2\right)}\right) \quad \text{and} \quad \text{the} \quad \text{solar} \quad \text{mass} \quad \text{to}$$

$$M_n = M\left(1 - \frac{GM}{2c^2 a}\right), \text{ we get, at last,}$$

$$\left(\frac{\partial r}{\partial \tau}\right)^2 = \frac{2GM_n}{r} - \frac{J_n^2}{r^2} + \frac{2GM_n J_n^2}{c^2 r^3} - \frac{GM_n}{c^2 a} - \frac{GM}{4c^2 a}$$

$$\left(\frac{\partial r}{\partial \tau}\right)^2 = \frac{2GM_n}{r} - \frac{J_n^2}{r^2} + \frac{2GM_n J_n^2}{c^2 r^3} - C_2 \quad (13.15).$$

Proceeding now exactly with the treatment used by Callahan (see Reference 8 - James J. Callahan, *"The Geometry of Spacetime"*, Springer-Verlag, Berlin (1999), pp 421 – 429) and by other authors, I make the following replacements:

$$r = \frac{1}{u}$$

$$\frac{\partial r}{\partial \tau} = -J\frac{\partial u}{\partial \theta},$$

So that equation (13.15) becomes

$$J_n^2\left(\frac{\partial u}{\partial \theta}\right)^2 + J_n^2 u^2 = 2GM_n u + \frac{2GM_n J_n^2 u^3}{c^2} - C_2 \quad (13.16)$$

Dividing by J_n^2 , differentiating with respect to θ ,and then dividing by $\dfrac{\partial u}{\partial \theta}$, one gets,

$$\left(\frac{\partial u}{\partial \theta}\right)^2 + u^2 = \frac{2GM_n u}{J_n^2} + \frac{2GM_n u^3}{c^2} - C_2$$

$$2\frac{\partial u}{\partial \theta}\frac{\partial^2 u}{\partial \theta^2} + 2u\frac{\partial u}{\partial \theta} = \frac{2GM_n}{J_n^2}\frac{\partial u}{\partial \theta} + \frac{6GM_n u^2}{c^2}\frac{\partial u}{\partial \theta} \qquad (13.17)$$

$$\frac{\partial^2 u}{\partial \theta^2} + u = \frac{GM_n}{J_n^2} + \frac{3GM_n u^2}{c^2}$$

There is no need to go any farther; equation (13.17) yields, except for the slightly modified angular momentum, J_n and the renormalized solar mass, M_n, essentially the same perihelion drift for Mercury as that derived by Albert Einstein in 1916 from the General Relativity Theory. The last term of equation (13.17) is a correction term from the SRVAR and it leads to the correct perihelion drift of the orbit. (The alert reader may have noticed that this term comes from the next to the last term of equation (13.16) which is simply the square of the escape velocity, at the point $\overline{r}$, of mercury from the gravitational field of the sun times the square of its tangential velocity, $\dfrac{J^2}{c^2 r^2}$, at the same point, $\overline{r}$. I will have a little more to say about this at the end of this section.)

None the less, I have not yet been able completely to convince myself that the derivation of equation (13.17) here in this work is completely correct. One could raise objections to the use of the SRVAR because of its peculiar mathematical irreversibility, which I have already mentioned. The SRVAR, for another thing, involves questions of mass and spin, which we do not yet fully understand. If use of the SRVAR should be found to be correct, then somehow space curvature has been accidentally introduced or else perihelion drift may not be indicative of space curvature (velocity curvature would, however, still be a possibility). To say the least, this result is astonishing.

The next example is central force motion of light in an open (hyperbolic) orbit. I consider a photon that has moved from infinity to a point at a distance $|\bar{r}| = R$ from the center of the sun, where the photon just grazes the surface of the sun and is traveling at a right angle to the radius vector, $\bar{r}$. A plane of symmetry, perpendicular to the orbit, passes through this point and the center of the sun. Now the motion of light is a bit different from that of a non-zero rest mass object. First the energy of the system is calculated using classical methods as though the photon is a point mass, with mass designated by, m_p, moving in the sun's gravitational field. The energy is calculated exactly at the point of symmetry of the orbit, thus,

$$\frac{m_p}{2}\left(v_r^2 + v_\theta^2\right) = \frac{GMm_p}{R} + \frac{m_p}{2}v_\theta^2$$

(13.18)

$$\frac{m_p}{2}v_r^2 = \frac{GMm_p}{R}$$

In order to calculate the momentum of the photon in the negative $\bar{r}$ direction, one must utilize the energy momentum equation from special relativity for a zero-rest mass particle, namely: $0 = -\mathcal{E}^2 + c^2 p^2$. Consequently, $p_r = \dfrac{\mathcal{E}}{c} = \dfrac{GMm_p}{cR}$. Now the momentum in the $\hat{\theta}$ direction is simply, $p_\theta = m_p c$, so that the half angle of deflection of the photon is given by :

$$\tan\left(\frac{\phi}{2}\right) \cong \frac{\phi}{2} = \frac{p_r}{p_\theta} = \frac{GMm_p}{m_p c^2 R} = \frac{GM}{c^2 R} \quad (13.19)$$

The total deflection for the photon is then

$$\phi = \frac{2GM}{c^2 R} \quad (13.20)$$

This is the so-called Newtonian result obtained also by other researchers. However, if equation (13.18) for the energy is modified to take into account the SRVAR; that is by including the velocity cross product term, the correct result is obtained. Thus,

$$\frac{m_p}{2}\left(v_r'^2 + v_\theta^2 - \frac{v_r'^2 v_\theta^2}{c^2}\right) = \frac{GMm_p}{R} + \frac{m_p}{2}v_\theta^2$$

(13.21)

$$\frac{m_p}{2}\left(v_r'^2 - \frac{v_r'^2 v_\theta^2}{c^2}\right) = \frac{GMm_p}{R}$$

Inserting into the cross product term of equation (13.21), the approximation for $v_r'^2$ coming from the second part of equation (13.18), that is from $\frac{m_p}{2}v_r'^2 \cong \frac{GMm_p}{R}$, and setting $v_\theta = c$, one finally gets,

$$\frac{m_p v_r'^2}{2} = \frac{GMm_p}{R} + \frac{GMm_p}{R}$$

(13.22)

$$\mathcal{E}' = \frac{2GMm_p}{R}$$

And therefore, $p_{r'} = \frac{\mathcal{E}'}{c} = \frac{2GMm_p}{cR}$, so that

$\tan\left(\frac{\phi'}{2}\right) \cong \frac{\phi'}{2} = \frac{p_{r'}}{p_\theta} = \frac{2GMm_p}{m_p c^2 R} = \frac{2GM}{c^2 R}$. Consequently, the total

deflection of the light by the sun's gravitational field is then given by,

$$\phi' = \frac{4GM}{c^2 R} \qquad (13.23)$$

Hence, one sees, once again, that application of the SRVAR yields the same result as the General Relativity Theory.

The third and last example is motion of a photon upward along a radius vector through the center of the earth from a point very near the surface of the earth to a point a few hundred meters, say, above the first point. The energy of the photon at the two points is compared. At the first point the potential energy of the gravitational field is $\frac{GMm_p}{r_1}$. If the second point is a height, h, above the first, then the potential energy is $\dfrac{GMm_p}{r_1 + h} \cong \dfrac{GMm_p}{r_1}\left(1 - \dfrac{h}{r_1}\right)$. Of course the

gravitation acceleration, g , is approximately constant over the

small (compared with r_1) distance h and $g = \dfrac{GM}{r_1^2}$, so that

$\hbar\omega_1 = \dfrac{GMm_p}{r_1}$ and $\hbar\omega_2 = \hbar\omega_1 - m_p hg$. But $m_p \cong \dfrac{\hbar\omega_1}{c^2}$ so that in the

end the result is $\hbar\omega_2 = \hbar\omega_1\left(1 - \dfrac{hg}{c^2}\right)$. Notice that the direction of the

velocity is constant, in a straight line, i.e. $v_r = c$ and $v_\theta = v_\phi = 0$ so
that the SRVAR correction is zero. Nevertheless, the correct result
was obtained the same as with General Relativity Theory.

This presents an awkward situation. It is not known that the
SRVAR involves space curvature, based as it is on a Minkowski
space. Involvement of space curvature seems unlikely. At most,
something like velocity curvature may be operative. At this time I
do not know for certain.

As I mentioned earlier in this section, the perihelion drift
correction term is equal to the square of the escape velocity times the
square of the tangential velocity. But this same factor, one can see is
involved in the bending of starlight (equation (13.22) and the
gravitational redshift where both the correction term and the
tangential velocity are zero. I mention also that the escape velocity
is independent of any spatial direction as is time which I take to be in
the direction given by the unit vector $\hat{l}$.

In light of the results in the previous paragraphs that cast doubt
on the experimental support for space curvature (as a basis for
gravitational force), namely the strong possibility that the perihelion
drift of the planet Mercury, the bending of starlight, the gravitational
redshift, and the Lense-Thirring effect can all be explained by a
proper version of the Special Theory of Relativity which
incorporates a correct velocity addition rule (and very possibly
gravitomagnetic effects), a new, perhaps, version of gravity is
proposed, pending independent confirmation or negation by other
physicists.

I begin by defining a gravitational charge, q_m ,

$$q_m = i\sqrt{G}M \quad (13.24)$$

This expression allows for the possibility of a negative gravitational charge if the source mass, M , is negative. The field potential energy 4-vector for an elementary particle of rest mass M is defined in analogy to that for the Electromagnetic field as

$$iV\hat{\sigma} = i\phi\hat{I} + \phi\bar{\bar{\beta}} = i\frac{q_m}{r}\hat{I} + \frac{q_m}{r}\bar{\bar{\beta}} \quad (13.25)$$

Operation of the differential $\square$ matrix on this vector gives

$$D(iV) = \begin{pmatrix} -\dfrac{i}{c}\dfrac{\partial}{\partial t} & \nabla \\ -\nabla & -\dfrac{i}{c}\dfrac{\partial}{\partial t} \end{pmatrix} \begin{pmatrix} i\phi\hat{I} \\ \bar{A} \end{pmatrix} = \begin{pmatrix} \left(\dfrac{1}{c}\dfrac{\partial\phi}{\partial t} + \nabla\square\bar{A} \right)\hat{I} \\ i\left(-\nabla\phi - \dfrac{1}{c}\dfrac{\partial\bar{A}}{\partial t} \right) + \nabla\times\bar{A} \end{pmatrix} \quad (13.26)$$

From this we see that the static gravitational force, $\bar{F}$, for particles of rest masses, M and m , is

$$\bar{F} = \frac{q_M q_m}{r^2}\hat{r} = \frac{i\sqrt{G}M\left(i\sqrt{G}m\right)}{r^2}\hat{r} = -\frac{GMm}{r^2}\hat{r}, \quad (13.27) \quad \text{which}$$

is, of course, Newton's law. From this and the definition of gravitational charge above, we see that the force is always attractive between positive masses. We also see that if one of the masses is negative, then the force is repulsive and, of course, if both are negative the force is again attractive. Consequently, any negative masses, if they exist, would by now be on the other side of the universe from us. If a negative mass were to collide with a positive mass they both should disappear in a cloud of nothing. No negative masses or energy have ever been detected, at least in our part of the universe and ten billion years or so after the big bang they should be "long gone". The absence of negative gravitational charge makes it impossible, or at least impractical, to shield against gravity, so Einstein's elevator thought experiments remain valid.

However, now we obtain a gravitomagnetic force field, $\bar{B} = \nabla\times\bar{A}$, but the force due to this field will be extremely weak. For example, at the surface of the Earth the gravitomagnetic force due to its rotation would be much too small to be detected by a gravimeter. Nevertheless, the gravitomagnetic field along with the normal gravitational field would give rise to gravitons emitted by accelerated masses. Because no negative masses are present, this would be quadrupole type radiation, which would require the gravitons to have spin equal to 2 and make the weak radiated field even weaker. It is expected therefore, that gravitational radiation will be difficult to detect directly.

Utilization of the techniques described in Chapter VI for the electromagnetic fields, enables one to derive all of the pertinent field equations for the gravitational fields from equations(13.24), (13.25), (13.26) and (13.27). Apart from the lack of any negative charges, the results of this theory are quite similar to Electromagnetism. One expects, for example that Black Holes and gravitons will exist. It is, perhaps, really not necessary for there to be any space curvature to have a gravitational theory that accords with all the now known facts. It seems reasonable then to apply Ockham's razor to space curvature, whilst possibly retaining velocity curvature in the explanation of gravitation.

In conclusion, turning to the larger question of what is the fundamental basis of the results presented here, one can begin to discern an underlying theme or logos (in the philosophical sense). A particular matrix operator which has its own algebra and other suitable mathematical properties has been used to determine the basic differential equations of a number of important areas of physics, i.e. of electromagnetic theory, classical mechanics, quantum mechanics, and relativity. It seems that all of the equations of each area are correctly obtained merely by application of a suitable operation without the necessity of applying the usual logic in reaching each step. The procedure or logos, which may not correspond to simple logic as now understood, depends mainly on the algebra of the matrices, which may appear to be "insane", but which yields the correct result. At this point, I can only say that simple logic is not everything.

The author hopes that a door has been opened by this work that leads to greater understanding of physics. While some significant discoveries have so far been made in the present work, it is clear that much more work remains to be done.

INDEX

A

D

E

117

M

Relativistic energy-momentum 4-vector   34, **50**, 54-58, 67, 79, 86, 95, 103, 105
Relativistic orbital equation   106-110
Rest momentum   34, 40
Retardation factor   46
Rotation matrices
        4-dimensional   18-20
        8-dimensional   20-23

S

Space curvature   110, 113, 115
Special relativity velocity addition rule (SRVAR)   32-33, 72, 73, 77, 78, 86, 90, 101, 105, 108, 110, 111, 112, 113
Spin   54-58, 64, 75, 77, 81-82, 85, **86-89,** 90-93, 95, 98, 101, 102, 104, 110, 115
Spin model for electron   86-94
Spin angular momentum   86
Spin force, time component of   81, 87-88
Symmetric periodic matrices   **3-4**, 44

T

Time averaged spin rotation matrices   92-93
Time component of spin force, F_4 81, 87-88

U

Unsnarling of 4-vectors   25-29
Unsnarling of momentum 4-vector   57-58
Unsquaring of 4-vectors   26, 32, 41-42

V

Velocity addition rule   **32-33**, 73
Velocity curvature   110, 113, 115
Velocity 4-vector   24, 100
Very generalized Lorentz transforms   67-68

W

Wave equation **51-54**, 64
Wavefunction   51-54, 78-81, 93-94, 95-98, 104
Wavefunction argument   53, 54, 104
Wavefunction renormalization   87
Wave-particle duality   48-53

X-Y-Z

X matrices   **3-4**, 44
Zero 4th component of E-M Field   38-41, 47, 103